THE
EVERYTHING.
EVERYDAY
MATH BOOK

D1296688

Dear Reader,

While I was writing this book, I heard a Wells Fargo advertisement on the radio that began with the following words: "There are numbers all around you. Your job is to determine what they mean." Those words exactly describe the intent of this book. Thinking and knowing about numbers has always been important for every single person, although most of the population would say that only scientists, accountants, and engineers use math. Living within a budget (or the consequences of not doing so), paying taxes (or the consequences of not doing so), and having enough materials to complete a home improvement project are practical, everyday applications of mathematics.

Because we use math every day, it is important to have a sense of what certain numbers mean. This is not to say that you have to be a human calculator. But when you hear a number used in describing a situation (unemployment rates, rate of inflation, final price after discount and taxes), isn't it helpful to be able to decide whether or not that number is reasonable? This book will help you improve your understanding of numbers and your comfort in working with them.

Chris Monahan

Welcome to the EVERYTHING® Series!

These handy, accessible books give you all you need to tackle a difficult project, gain a new hobby, comprehend a fascinating topic, prepare for an exam, or even brush up on something you learned back in school but have since forgotten.

You can choose to read an *Everything®* book from cover to cover or just pick out the information you want from our four useful boxes: e-facts, e-questions, e-alerts, and e-ssentials.

We give you everything you need to know on the subject, but throw in a lot of fun stuff along the way, too.

We now have more than 400 *Everything®* books in print, spanning such wide-ranging categories as weddings, pregnancy, cooking, music instruction, foreign language, crafts, pets, New Age, and so much more. When you're done reading them all, you can finally say you know *Everything®*!

QUESTION

Answers to common questions

FACT

Important snippets of information

ALERT

Urgent warnings

ESSENTIAL

Quick handy tips

PUBLISHER Karen Cooper

MANAGING EDITOR, EVERYTHING® SERIES Lisa Laing

COPY CHIEF Casey Ebert

ASSISTANT PRODUCTION EDITOR Alex Guarco

ACQUISITIONS EDITOR Pamela Wissman

DEVELOPMENT EDITOR Eileen Mullan

EVERYTHING® SERIES COVER DESIGNER Erin Alexander

Visit the entire Everything® series at *www.everything.com*

THE
EVERYTHING®
EVERYDAY
MATH BOOK

From tipping to taxes, all the real-world,
everyday math skills you need

Christopher Monahan

Former President, Association of Mathematics Teachers of New York State

adamsmedia
Avon, Massachusetts

Dedication: To all those who use math every day

Copyright © 2014 by F+W Media, Inc. All rights reserved.
This book, or parts thereof, may not be reproduced
in any form without permission from the publisher; exceptions
are made for brief excerpts used in published reviews.

An Everything® Series Book.
Everything® and everything.com® are registered trademarks of F+W Media, Inc.

Published by Adams Media, a division of F+W Media, Inc.
57 Littlefield Street, Avon, MA 02322 U.S.A.
www.adamsmedia.com

ISBN 10: 1-4405-6643-7
ISBN 13: 978-1-4405-6643-1
eISBN 10: 1-4405-6644-5
eISBN 13: 978-1-4405-6644-8

Printed in the United States of America.

10 9 8 7 6 5 4 3 2 1

Library of Congress Cataloging-in-Publication Data
Monahan, Christopher, author.
  The everything everyday math book : from tipping to taxes, all the real-world, everyday
math skills you need / Christopher Monahan.
    pages cm. -- (Everything series)
  Includes bibliographical references and index.
  ISBN-13: 978-1-4405-6643-1 (pbk. : alk. paper)
  ISBN-10: 1-4405-6643-7 (pbk. : alk. paper)
  ISBN-13: 978-1-4405-6644-8 (ebook)
  ISBN-10: 1-4405-6644-5 (ebook)
1. Mathematics--Popular works. I. Title. II. Title: Everyday math book.
  QA93.M625 2013
  640.1'51—dc23
                    2013030959

Many of the designations used by manufacturers and sellers to distinguish their products
are claimed as trademarks. Where those designations appear in this book and F+W Media
was aware of a trademark claim, the designations have been printed with initial capital
letters.

Cover images ©istockphoto/NickS, egunyeli; ©123RF/Maksym Yemelyanov; ©Veer/350jb.

This book is available at quantity discounts for bulk purchases.
For information, please call 1-800-289-0963.

Contents

Acknowledgments

I would like to thank Nick Galuardi, Mike Lapolla, and Erin Constantino, exercise physiologists from Saratoga Health and Wellness; Tracy Sherman of LaMarco Physical Therapy; Kellie Hummel of Franziska Racker Centers and the Tompkins County Department of Health Early Intervention Program; Gary Torrisi of Apple Pools; Dick Vaillancourt of Saratoga Springs Windows; Andrew Treanor of the Hendrick Automotive Group; Stacey Treanor, CFP®; and Tony, Steve, Mark, Debbie, Shaun, and Bruce of Vasyli for their help in writing this book.

Top Ten Times You'll Need to Use Math

1. After a nice dinner out with friends, and it's time to calculate the tip.

2. When you have to put enough gas in your car to drive a certain number of miles.

3. Just before you set the timer on your oven to cook Sunday dinner. How long does it take to cook a casserole, anyway?

4. When you only have $20 to spend on groceries.

5. When you're hosting a dinner party, and you need to triple the servings of a given recipe. How many cups is 6 tablespoons?

6. When you're trying to figure out what time to leave if it takes you 3½ hours to get to Grandma's house.

7. It's time for your morning run, but in order to burn 300 calories, how far will you need to go?

8. When you need to buy enough paint for two rooms.

9. When you need to pay the mortgage, car insurance, and electricity bill all in the same week.

10. Two words: tax season!

Introduction

JUST LIKE A PENDULUM swinging, the educational system in America has a history of fluctuating between teaching strategies. There are times when the American people and policy-makers feel that the liberal and performing arts must be emphasized more to produce kinder, gentler citizens. At other times, such as the end of the 1950s, the late 1970s, and the early twenty-first century, the emphasis on math and science is paramount. The problem with these swings is that a segment of the student population ends up discouraged, because they cannot make sense of the subject being emphasized. For example, in many high school math classes, it is not clear to the students how what they are learning is related to their everyday lives. This book is meant to help you better understand how much of the math you learned in math classes *does* apply to life outside the classroom.

I remember the first class my sophomore year at Manhattan College—a 9 A.M. literature class, twenty-one males sitting in the class (one math major and twenty engineers), and the professor announced that the course would be about American poets. You could hear the entire class gasp in unison. Not being one to get too involved in imagery, I assumed that my classmates reacted for the same reasons as I did. The professor then announced that we would read a different poet each class and then, at the end of the class, he would tell us what we needed to know about that poet for the exam. Well, that changed everything. It was a fun class, and I learned some interesting things that I might not have learned otherwise. No, I do not read poetry for pleasure or to gain insight or to be filled with admiration or moved to tears, but I do a little better when watching *Jeopardy!*.

I know I've had students sitting in my classes over the years who have had the same reaction to my math class as I had to that poetry class.

Calculators began appearing in the classroom in the late 1970s and early 1980s. At that time, it was not uncommon to hear a teacher say to a student, to a class, or to another teacher that the students had to learn

11

their addition and multiplication facts because they would not be carrying calculators with them all the time. How times have changed! Is there an electronic device that you carry in your pocket or purse that does *not* have a calculator built into it? Despite what your teachers may have said to you in the past, use it! Getting the right answer is still important, but getting the right answer quickly is just as important now. Was your teacher wrong? Well, yes and no. Ensuring that the answer (on your calculator) makes sense is an important skill you learned in school. All those estimating exercises that you did helped hone this skill. This is *the* key to using numbers effectively in the age of electronics. By all means let the machine do the work for you, but you must still be the one in control.

Whether you read this book to review arithmetic skills, to see how others use math on the job, or to gain confidence in your ability to fill out tax forms, you will see yourself in various places in this book. You are not alone in your struggle to feel more confident in your math skills. And if you have any doubt about the relevance of those skills to your life outside the classroom, this book will offer convincing evidence. It will illustrate many applications of math that you are likely to use every day.

CHAPTER 1

Are You a Math Phobe?

On October 4, 1957, the Soviet Union shocked the world by launching the first unmanned object into space, beginning what history would call the Space Race. When Americans saw what the Soviet Union had accomplished, their immediate conclusion was that American students were not being taught enough mathematics and science to keep up with worldwide competition. An innovative curriculum called New Math was instituted in American schools far and wide, but ironically, it created a generation of citizens who felt they had no understanding of mathematics or numbers. The major problem with New Math was that the teachers who were asked to teach the material were not given enough preparation. Teachers who were not specialists in mathematics were asked to teach some very sophisticated material, and the confusion they felt was passed on to their students. Although curriculums have changed since that time, the problem of not understanding math still exists. In fact, if you are still reading this book, there is a good chance that you or someone you know believes that you do not understand math.

What Is Math?

Simply put, mathematics is a way for people to describe the world. Many people also think that mathematics involves complex ideas and concepts. For example, some of the best-known mathematical theories—such as Newton's three Laws of Motion, Einstein's Theory of Relativity (which involves space, time, and gravity), and the Pythagorean theorem (which expresses the relationship among the three sides of a right triangle)—were created to explain how and why certain aspects of the universe exist. Those are big ideas! But math doesn't have to be that complicated.

You Use Math Every Day

Everyone uses much simpler mathematics on an everyday basis. A child knows that if he has two cookies and his sibling has three, there is a difference in how many they have. People who have tickets to a play in a city 50 miles away from where they live know that they can't leave their house only 20 minutes before the curtain rises if they want to make it to the show on time. These folks apply some sophisticated mathematics in deciding not only how long the drive should take, but also the probability of there being a traffic jam and how feasible it would be to use various alternative routes to get to the theater in time for the performance.

Take, for example, a trip to the supermarket. You use math when deciding what to buy. Consumers have used comparison shopping for years to make decisions based not only on price but also on quality. You probably consider all the facts before buying a product. How do product A and product B compare in cost? Will you eat more of product A than of product B, and thus get more nutrition from it, even if it is more expensive? Is it even worth it, in the long run, to buy product B instead of product A if you know you or your family won't eat it? Even though it is tempting to spend less, by purchasing product B rather than product A, the money is wasted if no one in the house will eat product B. All of these mathematical considerations play a part in your decision, whether you know it or not.

FACT

As of May 2013, nineteen states and two territories have unit pricing laws or regulations in effect.

What Is Math Phobia?

If you have no confidence in your ability to solve problems that involve numbers, or if you just defer to other people because they are "good at math," you are most likely a math phobe. What is a math phobe? When asked to respond to a math question, math phobes may have a physical reaction such as:

- Acute anxiety (exhibited as nervousness, pounding heart, rapid breathing, sweating, nausea, upset stomach, or tenseness)
- A feeling of panic or fear
- Fuzzy thinking
- A mental block in thinking

In extreme cases, the math phobe might even have feelings of helplessness, guilt, shame, inferiority, or stupidity.

Do you exhibit some of the characteristics described in the previous paragraph? If you answered yes, please ask yourself a few other questions: Do you know how much time you need to get from home to work, and the best way of getting there, depending on the time of day you leave your house? Do you know how to adjust a recipe so that you can make enough for the people who will be eating the meal instead of enough for the number of people the recipe is said to feed? Do you make financial decisions at work or at home? If you answered yes to any of these questions, you will have to say goodbye to the math phobe—and hello to the new you. You already *do* mathematics every day, so there is absolutely nothing to be afraid of! If hearing about a math problem that consists of multiplying two 2-digit numbers used to throw you for a loop, remember that while multiplying multi-digit numbers can be perplexing, you can use a calculator!

The Modern Computer and Calculator

In the mid-1980s, there was a great buzz in the computer industry about the newest product hitting the market. The desktop computer for the home, it was proclaimed, would make everyone's lives so much easier. In fact, computer designers told people, "Think of all the productivity and leisure time you will have because the computer will do so much for you!" They lied.

One result of the personal computer revolution has been that people now have more information to filter than ever before. Using the Internet to solve a problem or get an answer is second nature. But using a computer to seek a recipe for cooking chicken will yield more than a million search results. No wonder the computer intimidates so many people!

Thanks to those computing marketers 30 years ago, you now have tools to do math for you. Your responsibility is to make decisions that affect yourself, your family, and your job. Many times, those decisions are based on data that must be computed. Let the device do the work for you, whether it's in the form of a simple calculator, your desktop computer, your laptop, your smartphone, or an iPod, iPad, or other tablet.

Using a Calculator Is Not Cheating

Have you ever seen the 1980s movie *Wall Street*, with Michael Douglas? There is a scene in that movie where he uses his cellular phone while walking down the street. The phone in that scene relied on a battery pack that had to be carried in a bag with a shoulder strap because it was so large. The phone itself was about 10 inches long, and it weighed more than 2 pounds. The TI-10, one of the first Texas Instruments calculators (available in the mid-1970s) was about as big as that cell phone, and the battery life was so short that it was wise for the user to remain near an outlet in order to plug in the power pack while using the calculator. This calculator was capable of performing the four operations of addition, subtraction, multiplication, and division, and it had a percent key.

Technology has come a long way since then. Cell phones are more compact and have a wider array of abilities beyond making phone calls. Nearly all cell phones include a calculator that performs the same basic functions that the original TI-10 did, as well as an application to compute tips in restaurants. There are many basic calculators about the size of a credit card that are powered by solar cells. Whether you carry it in a pocket or a purse, you have a powerful tool at your disposal should you ever need to have an exact computation, as opposed to an approximation. Taking advantage of the tools available to you is always a good decision. It is not cheating!

FACT

The Texas Instruments calculator NSpire CX CAS has as much random access memory (RAM), 64k, as did the Apple II computer from the mid-1980s. The Nspire CX CAS is about the size of a cell phone.

Learn to Love Math

Try to learn to appreciate math. The old adage "Nothing succeeds like success" certainly applies to doing math, and there is a great deal of truth to it. Self-proclaimed math phobes may find that a few experiences of successfully solving a math problem alleviate their fear of the next problem that will come their way. The math phobe may become a "Math is okay" person with enough successes. And someday, maybe the math phobe will reach the "I love math" level!

You Can Help Others in Your Life with Math

Try to be a good math influence. There are few things more painful for a math teacher to hear from a parent than "I'm not surprised my child can't do math. I couldn't either." Avoiding negative statements about your skills or your spouse's skills will keep your children from developing preconceived notions about their own abilities. Positive statements go a long way toward giving children the confidence to take on a new task. Here are some further things you can do to help others develop confidence in their own abilities:

- Walk the talk. Try to convince others that what needs to be done *can* be done. As you find a solution, those watching you will notice your actions.
- Don't let them see you sweat. Naturally, you won't mention that you are uncomfortable calculating the tip, but also try not to show any of the physical signs (squirming, trembling, lack of eye contact) that might come with being uncomfortable.
- People learn from their mistakes. Everyone is a problem-solver; some people are just more adept than others at analyzing a problem quickly. Verbalizing the questions you ask yourself as you attempt to solve a problem has a number of wonderful effects. The observer is learning the art of problem-solving. Learning to ask questions gets you on the road to finding a solution. Even though there will be many times when the first questions asked do not lead to a solution, finding the right questions to ask is part of the learning process. Learning to evaluate a question and assess whether you can answer it is an important skill for everyone to have. Pass it along.

Math in the everyday world can be challenging for one big reason: There are too many steps! Even people with long attention spans and great abilities can get lost. Some basic problem-solving skills can help out:

1. Make a list. If your head is swimming with numbers, do a brain dump. In other words, list everything you know about the problem. Then cross off what you don't need.
2. Draw a picture. It doesn't take a Rembrandt to whip up a sketch of a room, and label its dimensions.
3. Make a table. Organizing the information so it makes more sense can point you to the solution.
4. Look for clues. Certain words will tell you what you need to do: *per* means "each," *squared* means "times itself," and even *is* means "equals" or "is equal to."
5. Rewrite the problem. Sometimes you just need to write things in a different way. And, sometimes writing an equation with variables is just the thing.
6. Check your answer. It is a good idea to ask, "Is this answer *reasonable*?"

QUESTION

When do you use math?
Before you continue reading this book, make a list of the ways that you use math. As you read through the book, you may be surprised as to the number of items you add to this list.

CHAPTER 2

Whole Numbers, Counting Numbers, and Integers

You began learning how to count with numbers when you were a child. Questions like "How many fingers am I holding up?" and "How old are you?" introduced the basics of math, while other questions like "Can I eat an entire cookie?" introduced more complex concepts such as whole numbers. Whether you realized it or not, these questions helped you understand the mathematical rules for whole numbers, which in turn set the stage for you to feel comfortable doing math with all kinds of numbers.

The Most Important Skill of All: Estimation

Two of the most important numbers in all of mathematics are 0 and 1. In this book, powers of 10 (10, 100, 1,000, etc.) will be added to the list of simple numbers that many people already know. As you work through the rest of this chapter, you will learn to estimate your answer as much as possible. Although using a calculator to get the correct answer to complicated calculations is highly encouraged, having a ballpark idea of the answer beforehand is critical. After all, how many times have you heard of someone forgetting to enter a 0, or typing a number wrong into the calculator, and swearing that the answer on the calculator screen must be correct because the machine does not make mistakes?

QUESTION

How Much Is That Zero Worth?
When estimating, zeros are a big deal. Remembering how to manipulate numbers with zeros is really helpful—and simple. *Rule: Thou shalt add zeros when multiplying by multiples of 10, 100, 1,000, and so on.* Suppose you have this problem: 4,000×80,000. To get the answer, find 4•8, and then add the zeros on the end. How many zeros? As many as in the original problem. 4,000×80,000=320,000,000 (there are 7 zeros in all).

When you divide with huge numbers that have lots of zeros, do the opposite of when you multiply—subtract zeros instead of adding them. Suppose you have this problem: 120,000÷6,000. Find 12÷6, and then add zeros on the end. How many? Subtract the number of zeros in the second number (the number you're dividing by) from the number of zeros in the first number (the number you're dividing into). That's how many zeros will be in your answer:

12 ÷ 6 = 2
120,000 ÷ 6,000 (subtract 3 from 4 to get 1 zero)
20

Determining How Reasonable a Number Is

A lot of everyday math relies on assessing the reasonableness of numbers. For example, when you hear that a person has an annual gross income of $50,000, does that match the statement that the person earns $15 per hour at his or her job? No. Someone who works 40 hours per week for 50 weeks is paid for approximately 2,000 hours annually. Thus you would expect a person who is paid $15 per hour to earn approximately 2,000 × $15 = $30,000 in a year, not $50,000.

However, it could be that the person works not 40 hours but 60 hours per week. Multiplying 60 hours per week by 50 weeks per year gives 3,000 hours annually. Multiplying this by $15 per hour gives a total of $45,000—a lot closer to $50,000 than to $30,000. In that case, the worker's estimate of an annual income of $50,000 is not unreasonable.

The skill of estimating an outcome depends on more than the arithmetic of the problem. An important consideration in estimation is the size of the numbers and of the answer. For example, if the hourly worker were being paid $14.78 per hour, rounding this wage to $15 would give a reasonable estimate of the annual income, though a little bit higher than the actual figure. On the other hand, if the hourly wage were $15.25, the estimate for annual income would be a little lower than the actual amount.

FACT

Rules for rounding: look at the number in the place to the right of the place value you are rounding to. If that number is 0, 1, 2, 3, or 4, make it and all numbers to the right of it 0. If the number is 5, 6, 7, 8, or 9, make it a zero as well as all the numbers to the right of it *and* increase the number to the left of it by 1. For example, when asked to round 83,741 to the nearest thousand, the place value to the right of the thousands is hundreds. There is a 7 in the hundreds place. Increase the 3 to a 4, make all the numbers to the right 0 and the rounded number is 84,000.

Another example comes from grocery shopping. Say you have decided that you will spend no more than $65 for groceries. When estimating the amount of the goods you are putting into your grocery cart, you might choose to round anything for which the cost ends in $0.25 or higher up to the nearest dollar (unlike what you were taught in school, where anything that ended in 5 or higher was rounded up to the next higher number). If you have items that cost $4.58, $9.89, $12.10, and $7.39, you might estimate that you will be spending $5 + $10 + $12 + $8 = $35 (compared to the actual value, $33.96). That approach to estimation will help you stay within your budget.

Basic Number Properties

There is an accepted order for how mixed calculations should be done in a problem. Upper elementary and middle school students learn the mnemonic "Please Excuse My Dear Aunt Sally" (or PEMDAS) to remember that the order in which calculations should be done is:

- **P**arentheses (or any grouping symbols) first.
- **E**xponents second.
- **M**ultiplication and **D**ivision have equal priority and are done from left to right as they appear in the problem.
- **A**ddition and **S**ubtraction have equal priority and are done from left to right as they appear in the problem.

Some examples for the order of operations are:

Compute: $83 + 4 \times 5 - 72 \div 2^3$

There are no grouping symbols in this problem—top priority has been examined. There is an exponent so 2^3 is the first item calculated. Recall that the exponent 3 tells you to multiply the number 2 three times. That is, $2^3 = 2 \times 2 \times 2 = 8$. The problem is now

$$83 + 4 \times 5 - 72 \div 8$$

Multiplication and division have equal importance and are computed from left to right as they appear in the problem. Calculate $4 \times 5 = 20$ and $72 \div 8 = 9$ and rewrite the problem as $82 + 20 - 9$.

Addition and subtraction also have equal priority and are executed from left to right as they appear in the problem. $82 + 20 = 102$ and $102 - 9 = 93$. Therefore, $83 + 4 \times 5 - 72 \div 2^3 = 93$.

Compute: $627 - 3(14 - 4^2 + 17)^2 \div (25 - 16)$

This appears to more complicated, but the order of operations will help you work through it. The first rule in the order of operations is to work with the parentheses. This tells you that you must calculate $14 - 4^2 + 17$ first. In this expression, exponents have top priority so compute $4^2 = 4 \times 4 = 16$. Subtract $14 - 16$ to get -2 and then add $-2 + 17$ to get 15. Since there is a second set of parentheses, calculate the value of the expression within it, $25 - 16 = 9$. The problem now becomes: $627 - 3(15)^2 \div 9$

Exponents have the top priority of all the remaining operations. Calculate $15^2 = 225$. The problem is now $627 - 3(225) \div 9$. The expression $3(225)$ is an implied multiplication statement (a consequence of writing multiplication problems in algebra when x is often used as a variable $- 3 \times x$ is difficult to read so the multiplication is written as $3x$ and this implied multiplication has carried over to problems written with constants). Therefore, $627 - 3(225) \div 9$ becomes $627 - 675 \div 9$. Division has the next highest priority so the problem becomes $627 - 75$. Subtract to get the final answer, 552.

There are three important properties of numbers you should know. The commutative property and the associative property are true for both multiplication and addition. The third property is the distributive property of multiplication over addition (it is more commonly called the distributive property).

The *commutative property* tells you that you can add or multiply in any order that you prefer without changing the answer. You might think of it as commuting—going or coming is not the issue, you will travel the same distance both ways. Consequently, $2 + 3$ and $3 + 2$ give the same answer (and so do 2×3 and 3×2).

The *associative property* works with more than two numbers and tells you that you can arrange numbers to fit your preference. For example, calculating $8 + 4 + 6$ does not require that you add 8 and 4 first. Adding 4 and 6 to get 10 and then adding 8 to get 18 might be more natural, but either way, you will get the same answer.

ESSENTIAL

Knowing the names of the properties is important only if you are working with a youngster who is studying these number properties in school.
Knowing *how the properties work* is what you should concentrate on most.

The *distributive property* of multiplication over addition will prove to be important to learning how to do some mental multiplication. This property says that when a number is multiplied by a sum of two other numbers, you can choose how to do the problem. You should generally choose to do it the easier way, and you can often complete the calculation "in your head." For example, $12 \times (10 + 3)$ can be accomplished by doing the problem in these steps: $12 \times 10 + 12 \times 3 = 120 + 36 = 156$. (Use a calculator if you wish to check that 12×13 is 156.) On the other hand, the problem $9 \times 7 + 9 \times 3$ is more easily done as $9 \times (7 + 3) = 9 \times 10 = 90$.

Adding Lists of Numbers

Adding multidigit numbers in your head can be unnerving, especially if there are other people with you and they are waiting for you to give them an answer. Depending on the situation, estimating the answer may be appropriate. If so, apply the rules for rounding (if the number to the immediate right of the place you are rounding to is 0, 1, 2, 3, or 4, leave the digit alone; if that number is 5 or higher, increase the digit to the next higher number), and give an answer. For example, to estimate the sum of $157 + 496 + 814$, you might rewrite the problem as $200 + 500 + 800$. Next reorder the problem (as the commutative property permits) to $200 + 800 + 500$ and add $2 + 8 + 5$ to get 15. (You added $2 + 8$ to get 10 and then 5 more to get 15, right?) So your estimate is 1,500.

ESSENTIAL

The number directly to the left of the decimal point is the units (or ones) place. As you move to the left, each place value is 10 times the previous value. In the number 45,678, 8 is in the units place, 7 is in the tens place, 6 is in the hundreds place, 5 is in the thousands place, and 4 is in the ten thousands place. Numbers to the right of the decimal point are one-tenth the value of the place to the left of it. The "th" at the end of the place value indicates that the value is smaller than one. In the number 1.2345, the 1 is in the units place, the 2 is in the tenths place, the 3 is in the hundreths place, the 4 is in the thousandths place, and the 5 is in the ten thousandths place.

If estimating to the nearest hundred is not appropriate, you can estimate the problem to the tens place. The sum $157 + 496 + 814$ then becomes $160 + 500 + 810$. The ones place is zero, so make a mental note of that. The tens place adds to $6 + 0 + 1 = 7$. The hundreds place adds to $1 + 5 + 8$. If adding these numbers to get 14 works for you, then you have the estimate 1,470.

If you are not comfortable with the addition of $1 + 5 + 8$, don't worry—just break down the problem. This is a great example of the importance of the number 10. You know that $5 = 3 + 2$. If you rewrite $1 + 5 + 8$ as $1 + 3 + 2 + 8$, the problem becomes $4 + 10$, or 14. Breaking down the numbers so that you can get sums of 10 gives you a safe and reasonably quick way of adding more than two numbers. The key is to keep it simple. Don't try to add eight numbers at a time. Break the problem down into simpler steps.

Using Columns

If there is a need for the exact value of the sum of $157 + 496 + 814$, you could do it the way you probably learned in elementary school. That is, write the numbers in a column:

$$
\begin{array}{r}
157 \\
496 \\
+\ \underline{814}
\end{array}
$$

Start with the ones place. $7 + 4 + 6 = 7 + 10 = 17$. Write the 7 down, but since you now have a new 10 in the mix, carry the 1 to the tens column.

$$
\begin{array}{r}
1^{1}57 \\
496 \\
+\ \underline{814} \\
7
\end{array}
$$

Add the tens values, starting with this new value in the tens column: $1 + 5 + 9 + 1 = 6 + 10 = 16$. Sixteen tens is the same as 160. Write the 6 down, and carry the 1 to the hundreds column.

$$
\begin{array}{r}
{}^{1}1^{1}57 \\
496 \\
+\ \underline{814} \\
67
\end{array}
$$

Add the hundreds values: $1 + 1 + 4 + 8 = 14$. The final answer is 1,467.

^{1}1^{1}57
 496
+ 814
1,467

This is the most common approach to teaching addition with multiple digits.

Now let's try to solve this problem with mental math. In essence, the original problem of $157 + 496 + 814$ can be viewed as $100 + 50 + 7 + 400 + 90 + 6 + 800 + 10 + 4$ when you break the problem into place value amounts. When you rearrange these numbers (remember the commutative property), the problem looks like this:

$$100 + 50 + 7 + 400 + 90 + 6 + 800 + 10 + 4 = 100 + 400 + 800 + 50 + 90 + 10 + 7 + 6 + 4$$

Add the hundreds together, the tens together, and the ones together:

$$(100 + 400 + 800) + (50 + 90 + 10) + (7 + 6 + 4) = 1300 + 150 + 17$$

Adding the amounts in the ones place gives 7, adding the amounts in the tens place gives 6, adding the amounts in the hundreds place gives 4, and there is a 1 in the thousands place. The sum is 1,467. Is your head spinning? It might very well be. Here is a wonderful opportunity to repeat an earlier statement: Using a calculator is not cheating. If you need an exact answer, use the calculator that is on your computer or cell phone to get it.

Subtracting

Although the process for adding numbers hasn't changed much over the decades, the way in which subtraction has been taught and learned has undergone many variations. Before looking at these processes (which are also called algorithms), it will be worthwhile to learn a trick that will make some of the mental subtractions easier to do.

First, subtract $14 - 9$. If you can look at that problem and just say 5, go to the next paragraph. If you hesitated when you saw $14 - 9$, try this: $9 = 10 - 1$. So to subtract 9 from 14, you could subtract 10 from 14 to get 4. Because you subtracted 1 more than you should have, the real answer to the problem must be 1 more than 4, or 5. (The mathematical way to write this would be $14 - 9 = 14 - (10 - 1) = 14 - 10 + 1 = 4 + 1 = 5$.)

Consider the problem $842 - 658$. Let's look at three "different" approaches for finding the answer to this problem.

The first way to solve this problem is called *borrowing*. It sometimes helps to talk out the problem. In this case, the dialogue might sound something like this: "Since you can't subtract 8 from 2, you will have to borrow a 10 from the 4 and make the 2 a 12."

$$8^3 4^1 2$$
$$-\underline{6\ 5\ 8}$$

"12 minus 8 is 4."

$$^7 8^{13} 4^1 2$$
$$-\underline{6\ 5\ 8}$$
$$4$$

"You cannot subtract 5 from 3, so borrow 1 from the 8 and make the 3 a 13. 13 minus 5 is 8."

$$^7 8^{13} 4^1 2$$
$$-\underline{6\ 5\ 8}$$
$$8\ 4$$

"Finally, $7 - 6$ is 1. So the answer is 184."

$$^7 8^{13} 4^1 2$$
$$-\underline{6\ 5\ 8}$$
$$1\ 8\ 4$$

The second approach is really the same thing as borrowing, but it is called *regrouping*. For this example, the dialogue may sound like this: "You cannot subtract 8 from 2, so instead you should break one 10 from the 40 in 842 and rewrite it as 830 and 12."

$$\begin{array}{r} 842 \\ -\underline{658} \end{array} \qquad \begin{array}{r} 83{}^{1}2 \\ -\underline{65\ 8} \end{array}$$

"12 minus 8 is 4. You cannot subtract 5 from 3, so you should borrow 100 from the 800 and rewrite 830 as follows":

$$\begin{array}{r} 83{}^{1}2 \\ -\underline{65\ 8} \\ 4 \end{array} \qquad \begin{array}{r} 7{}^{1}3{}^{1}2 \\ -\underline{6\ 5\ 8} \\ 4 \end{array}$$

"13 minus 5 is 8, and 7 minus 6 is 1, so the answer is 184."

$$\begin{array}{r} 83{}^{1}2 \\ -\underline{65\ 8} \\ 1\ 8\ 4 \end{array} \qquad \begin{array}{r} 7{}^{1}3{}^{1}2 \\ -\underline{6\ 5\ 8} \\ 1\ 8\ 4 \end{array}$$

Both of these methods address place value and how these values are used to make the subtraction work properly. The third method is a relatively new application that tackles the issue of place value first and does not concern itself with the intricacies of borrowing or regrouping.

FACT

The number being subtracted is the subtrahend, and the number it is being subtracted from is the minuend. Again, these words are worth concerning yourself about only if you are working with a youngster who has encountered them.

The third process takes the number being subtracted, 658, and deals with it in place value form, which is also referred to as expanded notation: $600 + 50 + 8$. The problem starts by subtracting 600 from 842.

$$\begin{array}{r} 842 \\ -\underline{600} \\ 242 \end{array}$$

Now subtract 50 from 242.

$$\begin{array}{r} 842 \\ -\underline{600} \\ 242 \\ -\underline{50} \end{array}$$

Subtracting $2 - 0$ is easy enough, but what about $24 - 5$? First try counting down from 24: 23, 22, 21, 20, 19 (that is, $24 - 1 = 23$, $24 - 2 = 22$, and so on, until you reach $24 - 5 = 19$).

$$\begin{array}{r} 842 \\ -\underline{600} \\ 242 \\ -\underline{50} \\ 192 \end{array}$$

Then subtract 8 from 192 in a similar manner.

$$\begin{array}{r} 842 \\ -\underline{600} \\ 242 \\ -\underline{50} \\ 192 \\ -\underline{8} \\ 184 \end{array}$$

For the purposes of mental subtraction, this process actually works fairly well.

ALERT

Estimating with subtraction can be tricky. When estimated to the nearest thousand, the difference between $4367 - 2519$ would be $4000 - 3000 = 1000$. The impact of increasing one of the values and decreasing the other will exaggerate the difference (making it either too large or too small). Rounding both numbers down or both numbers up will give a better estimate. For example, $4367 - 2519$, when both are rounded down, becomes $4000 - 2000 = 2000$.

Try this example: Subtract 2519 from 4367.

$$
\begin{array}{r}
4367 \\
-\underline{2000} \\
2367 \\
-\underline{500} \\
1867 \\
-\underline{10} \\
1857 \\
-\underline{9} \\
1848
\end{array}
$$

Multiplication

There are a few tricks you can use to help with mental multiplication. The two most important tricks involve the results of multiplying by 0 (the answer is always 0) and multiplying by 1 (the answer is always the number by which the 1 is multiplied). When estimating products, try to round at least one of the numbers to a multiple of 10 for two-digit numbers, to a multiple of 100 for three-digit numbers, and so on. The rule for multiplying whole numbers ending in zero is that the product has as many zeros as the factors had.

FACT

The numbers being multiplied are called *factors*. The answer to the multiplication is called the product.

A multiplication table can help with learning the basic facts.

MULTIPLICATION TABLE

×	2	3	4	5	6	7	8	9
2	4	6	8	10	12	14	16	18
3	6	9	12	15	18	21	24	27
4	8	12	16	20	24	28	32	36
5	10	15	20	25	30	35	40	45
6	12	18	24	30	36	42	48	54
7	14	21	28	35	42	49	56	63
8	16	24	32	40	48	56	64	72
9	18	27	36	45	54	63	72	81

To multiply two numbers with this table, move down the left column until you find your first number (the first factor in the problem). Then slide across this row until you are under the column headed by the second factor. For example, to multiply 4×7, move down the left column to 4, and then read across the row until you are under the column header 7 to get the product 28.

If you wish, you can move across the top row to the number 4 and down that column to the row beginning with 7, and you will get the same answer. You have just illustrated the beauty of the commutative property. (But note that not all tables work this way. For example, if you were to make a table for subtraction, you would have no choice about which number to use first.)

Why is the number 1 missing from the table? Well, multiplying by 1 always gives an answer identical to the other factor in the problem (the number that is multiplied by 1). For example, $3 \times 1 = 3$. Practice as much as you need to with this table to be sure that the results are very familiar to you. Committing it to memory is a good idea, because let's face it, learning your "times tables" can save you a lot of time!

Here is another problem to solve: Estimate the product of 329×487.

First, round the numbers to 300 and 500. There are a total of four zeros in the factors, and $3 \times 5 = 15$. The estimate for 329×487 is 150,000. (If you need the exact answer of 160,023, use your calculator.)

The distributive property plays an important part in most of the multiplication shortcuts. As a reminder, the product $8 \times (7 + 5) = 8 \times 7 + 8 \times 5 = 56 + 40 = 96$.

Multiplying by 15

Because $15 = 10 + 5$, you can easily multiply any number by 10 just by adding a zero to the end of the first factor. To multiply by 5, take half the product that you found when you multiplied by 10. Add these results together. For example, 24×15 becomes $24 \times (10 + 5)$. Next $24 \times 10 = 240$, and half of that is 120. So $24 \times 15 = 240 + 120 = 360$.

Here's another example: 37×15

37×15 becomes $37 \times (10 + 5)$. $37 \times 10 = 370$. Half of 370 might not roll off your tongue that quickly, but practice will tell you it is 185.

$370 + 185 = 300 + 100 + 70 + 80 + 5$ (from the section on addition) $= 400 + 150 + 5 = 555$

Squaring Numbers That End in 5

The process for squaring a number ending in 5 is relatively straightforward, and learning it will prove helpful with other exercises in mental multiplication. Examine the following examples:

$$15^2 = 225, 25^2 = 625, 35^2 = 1225, 45^2 = 2025, \text{ and } 55^2 = 3025$$

What do all these answers have in common? For one thing, they all end in 25. Look at the number before the 5 in the factor (the number to the left of the equation sign) and the number before the 25 in the product (the number to the right of the equation sign), and you should see a pattern emerge. For 15^2, the number in the tens place is 1, but the number in front of 25 in the product is 2. For 25^2, there are 2 and 6. For 35^2, there are 3 and 12. Notice that $1 \times (1 + 1) = 2$; $2 \times (2 + 1) = 6$; $3 \times (3 + 1) = 12$. Does the pattern continue? $4 \times (4 + 1) = 20$ and $45^2 = 2025$.

To square a number ending in 5, write 25 at the end of the product. Then add, before that 25, the product of the number in front of the 5 in the original factor and one more than this number. Using a variation on mathematic notation, this means that $(b5)^2 = [b \times (b + 1)]25$.

Example: $65^2 = [6 \times 7]25 = 4,225$
$125^2 = [12 \times 13]25 = 15,625$ (Check that with your calculator.)

ESSENTIAL

Squaring a number means multiplying a number by itself. The exponent 2 is used to represent squaring. For example, $5^2 = 5 \times 5 = 25$.

Multiplying Two-Digit Numbers

A trick for mentally multiplying two-digit numbers is to rewrite at least one of the factors as a sum or difference of a multiple of a convenient number, and then to apply the distributive property. For example, to find the product 19×12, you could write 19 as $20 - 1$, and the problem would then become $12 \times (20 - 1) = 12 \times 20 - 12 \times 1$.

Next, $12 \times 20 = 240$ and $12 \times 1 = 12$. The difference $240 - 12$ can be found by doing $240 - 10 - 2 = 230 - 2 = 228$. Could the problem have been done by rewriting 12 as $10 + 2$? Yes, it could have. $19 \times 12 = 19 \times (10 + 2) = 19 \times 10 + 19 \times 2$. $19 \times 10 = 190$ is easy enough. How comfortable are you multiplying 19 and 2? When you have the option of changing either number, leave the smaller digits alone, because they are easier to multiply.

$$32 \times 97 = 32 \times (100 - 3) = 32 \times 100 - 32 \times 3 = 3200 - 96 = 3200 - 90 - 6 = 3110 - 6 = 3104$$

Now find the product 16×14. There are three different methods for doing this problem because the numbers are "special."

First option: rewrite $16 \times 14 = 16 \times (10 + 4) = 160 + 16 \times 4 = 160 + 64 = 224$.

Second, because 16 is one more than 15, the problem could be written as $(15 + 1) \times 14 = 15 \times 14 + 1 \times 14$. $15 \times 14 = (10 + 5) \times 14 = 140 + 70 = 210$. Add 14 (from 1×14) to 210 to get 224.

The third option involves a formula learned in algebra called the difference of squares. In algebra, letters (called variables) are used to represent numbers. With a and b representing different numbers, the formula reads as follows: $(a + b) \times (a - b) = a^2 - b^2$.

Because $16 = 15 + 1$ and $14 = 15 - 1$, this formula would apply, so $16 \times 14 = (15 + 1) \times (15 - 1) = 15^2 - 1^2 = 225 - 1 = 224$

The difference-of-squares formula takes a lot more practice to understand and use than the distributive property with multiples of 10. But it will be a valuable tool when you get accustomed to using it.

Multiplying by 25

You may have learned in school that a trick for multiplying by 25 is to multiply by 100 and then divide by 4. For example, 25×32 is the same as $3200 / 4 = 800$. It works. If that does not resonate with you, ask yourself this question: "How much money is 32 quarters worth?" You will probably say that 4 quarters make a dollar, so with 8 sets of 4, the answer is $8.00. You'd be surprised how much more accurate you can be when you think of the numbers in terms of simple money matters.

Division

You go out to dinner with 3 other people with the understanding that the bill will be divided equally. When dinner is over and the bill is received, a tip is added and the total is $177.89. To estimate the amount each person owes, divide 4 into $180, and you will get $45 each. How you choose to handle the difference between $180 and $177.89 is a personal issue.

Why divide 4 into 180? Knowing a few rules about divisibility can help you estimate numbers so that you can get a handle on the size of the values involved.

BASIC RULES FOR DIVISIBILITY

Divisible by	Property
2	The number ends in 0, 2, 4, 6, or 8.
3	The sum of the numbers is divisible by 3.
4	The last two digits in the number are divisible by 4.
5	The number ends in 0 or 5.
6	The number is divisible by both 2 and 3.
9	The sum of the numbers is divisible by 9.
10	The number ends in 0.

There are divisibility tests for a few other values, but they can be cumbersome to learn and use. In the dinner expense example, the total bill is estimated to be a value that is divisible by 4. 180 is close to 177.89 and still provides an easy number to divide. Half of 180 is 90, and half of 90 is 45.

Example: What are some of the divisors of 156? (The divisors of a number are the numbers that divide evenly into that number, with no remainder left over.)

156 ends in an even number, so it is divisible by 2. $1 + 5 + 6 = 12$, a number divisible by 3 (but not by 9), so 156 is divisible by 3 (but not by 9). 56 is divisible by 4, so 156 is divisible by 4.

Example: What are some of the divisors of 7,560?

The last two digits are 6 and 0, so you know that 7,560 is divisible by 4 (and also by 2). The number ends in a 0, so 7,560 is divisible by 10 (and by 5). $7 + 5 + 6 + 0 = 18$, so 7,560 is divisible by 9 (and by 3). You now know that 7,560 is divisible by 2, 3, 4, 5, 6, 9, and 10.

FACT

Because 7,560 is divisible by 4 and 9, does it make sense to you that 7,560 is divisible by 36? You may find yourself considering other combinations of the basic factors that lead to larger factors. Remember to be sure that the smaller numbers do not share common factors. For example, you would not claim that 7,560 is divisible by 50 because it is divisible by 5 and 10.

Negative Numbers

An integer is defined as any whole number (a nonfraction and nondecimal number) or the negative of a whole number. Adding and subtracting negative (signed) numbers usually makes sense to people, but multiplication and division often take a bit more explaining.

Addition

One way to think about negative (signed) numbers is to treat positive numbers as a gain and negative numbers as a loss. If you gain $100 and then lose $75, you are left with $25, so $100 + (-75) = 25$. If you gain $100 and then lose $125, you are left with a loss of $25, so $100 + (-125) = -25$. Losing $100 and then losing another $25 leaves you with a loss of $125, so $-100 + (-25) = -125$. (Parentheses are used when the negative number follows the addition sign, simply to emphasize the two symbols.) In general, when adding two numbers with the same sign, add the numbers and keep the sign. When adding two numbers with different signs, subtract the values and keep the sign of the larger value.

$$50 + (-45) = 5$$
$$70 + (-90) = -20$$
$$-30 + (-15) = -45$$

Subtraction

Thinking about subtracting signed numbers in the context of gains and losses can be confusing. The easiest way to think about subtraction with signed numbers is to view it as the opposite of addition. Consequently, you can change the subtraction to addition *and* change the sign of the second number. You may not have much trouble seeing that $6 - 10 = -4$, but $6 - (-10)$ is easier once you apply the rule you just learned: $6 - (-10) = 6 + (+10) = 16$.

$$-12 - 8 = -12 + (-8) = -20$$
$$-23 - (-18) = -23 + 18 = -5$$

ESSENTIAL

When entering a problem such as $-12 - (-8)$ into a four-function calculator, use the $\pm$ on the calculator to change signs. Type $12 \pm - 8 \pm =$ to get the answer, -4.

Multiplication

You know that $3 + 3 + 3 + 3 + 3$ is the same as 5×3 and is equal to 15. In the same way, $-3 + (-3) + (-3) + (-3) + (-3)$ is the same as $5 \times (-3) = -15$. Multiplying two numbers with different signs yields a negative answer, and the product of two negative numbers is a positive number. A good example to help you understand is $-7 \times 0 = 0$. Since 0 can be written in many different ways (for example, $-5 + 5$), the problem -7×0 can be rewritten as $-7 \times (-5 + 5)$. Use the distributive property to expand $-7 \times (-5 + 5)$ as $-7 \times (-5) + (-7) \times (5)$. Then -7×5 is -35. The number that is added to -35 to get 0 is 35. Therefore, it must be the case that $-7 \times (-5) = 35$.

Example: $-8 \times (-5) = 40$
Example: $12 \times (-15) = -180$

Division

Division and multiplication are opposite operations. If $12 \times (-15) = -180$, then $\dfrac{-180}{12} = -15$ and $\dfrac{-180}{-15} = 12$.

ESSENTIAL

When you multiply or divide two numbers with the same sign, the answer is positive; if the signs are different, the answer is negative.

$-200 \div (-25) = 8$

$400 \div (-16) = -25$

Negative Numbers and Spreadsheets

There are four settings on a spreadsheet that will treat negative numbers in different ways. In General mode, $25 - 40 = -15$. In Number or Currency mode, $25 - 40$ can be displayed as -15, (15), or 15 (but in a red font). In Accounting mode, $25 - 40$ will be displayed as (15).

QUESTION

When do you use integers?
Consider the activity of taking an inventory at work or home, deciding what to wear when you see the temperature as you leave for work versus the change in temperature when you will be returning home, whenever you perform an estimation (grocery bill, time to get to your destination, etc.). What else did you add to your list of when you use math?

Exercises

Estimate the results for questions 1 through 5 to the nearest hundred.

1. $246 + 892 + 367 + 739$
2. $1207 + 3268 + 4349 + 8237 + 832$
3. 83×28
4. $854 - 523$
5. $5708 \div 82$

Estimate the results for questions 6 through 8 to the nearest thousand.

6. $1207 + 3268 + 4349 + 8237 + 832$
7. 457×81
8. $8739 - 2308$

Perform each of the operations mentally, using some of the hints discussed in this chapter (or techniques you have developed on your own). Check each answer with a calculator.

9. $432 + 587 + 219$
10. $653 + 418 + 791$
11. $867 - 572$
12. $3743 - 1676$
13. 23×31
14. 95^2
15. 83×19
16. 47×43
17. 25×43
18. $-100 + (-40) + 210$
19. $-400 - 120$
20. $85 - (-75)$
21. $24 \times (-26)$
22. $-19 \times (-14)$
23. $-580 \div 29$
24. $-850 \div (-25)$
25. $2870 \div (-14)$

Make up as many more exercises as you'd like to practice. Write the problems down so that you can see them, do them mentally, and then check your answers with a calculator.

CHAPTER 3

Decimals

Writing parts of whole numbers involves the use of fractions and decimals. Since the arithmetic of decimals is so similar to the arithmetic of whole numbers, decimals will be discussed first in this chapter, and fractions will come next.

Place Value—Big Numbers and Small Numbers

According to the website *www.treasurydirect.gov*, on March 22, 2013, the national debt was $16,749,924,938,094.43. That is a lot of money. *Voyager 1*, a satellite that was launched from Earth in 1977, left our solar system in March 2013 and has traveled more than 11 billion miles. That is a long way to travel. Both of these numbers boggle the mind, first because of their size and second because of what they mean.

Another number to consider is the download speed from the Internet to your home computer. Say your download speed is 15.4 Mbps (15.4 million bits per second). That is a great deal of information coming into the computer each second, and yet, many people find that this is too slow. Nanotechnology is one of the growing industries in the United States today. Computer chips are becoming smaller in size, while also becoming greater in capacity and faster in transmitting information. A nanometer (0.000000001 meter) is less than 1/1000 of the width of a human hair. That is a small number.

Consider the impact that this technology has had in your recent life. MP3 players and smartphones are becoming smaller, with greater capacities and lower prices. It is now possible to buy televisions with 3-D capabilities. Will home theater systems using holography be available to the general public in the near future? One can only imagine.

FACT

Phone companies make claims about the download speeds of their 4G systems as compared to 3G systems. According to PCMag (*www.pcmag.com/article2/0,2817,2399984,00.asp*), this might not be the case. You should read the specifications for each phone that you are considering and determine which phone has the greater speed.

Scientific Notation

There is no question that scientific discoveries have vastly extended the scale and range of our ability to understand the world. In other words, "big" is growing vastly bigger, and "small" is getting tinier and tinier. But even though both extremes are beyond imagining, it is possible to quantify them so that they can be compared and used in calculations. In discussing numbers of the sizes mentioned in the preceding section, it is common to write them in what is called scientific notation. The rule of thumb is that the number is written in the form x.xx × 10n, where n is some positive or negative integer. The value of n is determined by how many decimal places the decimal point must be moved to get the value *x.xx*. If the decimal point is moved to the left, n is positive. If the decimal point is moved to the right, n is negative. For example, the national debt \$16,749,924,938,094.43 would be written as \$1.67 × 10^{13}, and a nanometer is 1 × 10^{-9} meter. *Voyager 1* has traveled more than 1.1 × 10^9 million miles.

FACT

Numbers written in scientific notation always have one digit to the left of the decimal point and usually two digits to the right of the decimal point.

The Metric System

The United States is one of the few nations that does not use the International System of Units (SI), better known as the metric system. But the metric system offers many benefits. It is based on scientific measurement, and more important, it uses the decimal system. So, if the United States adopted the metric system, then rather than comparing a ½-inch wrench to a ⅝-inch wrench, you would compare a 13-millimeter (mm) wrench to a 16-mm wrench, which is way easier.

To move from one level of magnitude to another within the metric system, you just multiply or divide by powers of 10. The prefixes most commonly used in the metric system are *kilo-*, *deci-*, *centi-*, and *milli-*. Thanks to the impact of the computer industry, the prefixes *tera-*, *giga-*, *mega-*, *micro-*, and *nano-* have also joined the everyday vocabulary of many people.

ESSENTIAL

One of the major advantages of the metric system is that it is based on the decimal system, making calculations and conversions within the system much easier to do. For instance, 4 kilometers can be converted to meters merely by multiplying by 1,000.

PREFIXES, SYMBOLS, AND FACTORS OF THE METRIC SYSTEM

Prefix	Symbol	Factor
tera	T	$1,000,000,000,000 = 10^{12}$
giga	G	$1,000,000,000 = 10^{9}$
mega	M	$1,000,000 = 10^{6}$
kilo	k	$1,000 = 10^{3}$
hector	h	$100 = 10^{2}$
deca	da	10
deci	d	$0.1 = 10^{-1}$
centi	c	$0.01 = 10^{-2}$
milli	m	$0.001 = 10^{-3}$
micro	µ	$0.000001 = 10^{-6}$
nano	n	$0.000000001 = 10^{-9}$
pico	p	$0.000000000001 = 10^{-12}$

The basic units in the metric system are meter for length, gram for mass, and liter for volume. One meter is approximately 39 inches (39.3701 inches) long. A kilometer (1 km = 1000 meters) is approximately ⅝ of a mile. It might be easier to remember that 5 miles is equivalent to 8 km. (Isn't it interesting that the racing community will run 5k and 10k races locally, but marathons are still 26.2 miles?) On a smaller scale, 1 inch = 2.54 cm = 25.4 mm.

FACT

A 2-liter bottle of soda, a very common sight in grocery stores and convenience stores, contains 67.628 ounces of soda. That is 3.628 ounces more than a half gallon. A gallon of gasoline is the equivalent of 3.78541 liters of gasoline.

The medical profession uses the term *cubic centimeter* when discussing the amount of fluid a patient is given. One cubic centimeter (cc) is equal to 1 milliliter (ml), so 1000 cc = 1 liter (L). One kilogram is the equivalent of 2.2 pounds. The packaging on food products reports the weight (or volume) of the product in both standard and metric values.

Temperature in the SI system is measured in degrees Celsius (or degrees centigrade), but Fahrenheit measurement is still used in the United States. The formula for converting degrees Fahrenheit to degrees Celsius is

$C = \frac{5}{9}(F - 32)$. And the formula for converting degrees Celsius to degrees Fahrenheit is $F = \frac{9}{5}(C + 32)$. A quick way to *estimate* the Celsius temperature from the Fahrenheit temperature is to use the formula $C = .5(F - 30)$. In other words, subtract 30 from the Fahrenheit temperature and then take half. A quick way to *estimate* the Fahrenheit temperature from the Celsius temperature is to use the formula $F = 2C + 30$. In other words, double the Celsius temperature and then add 30. Your estimate might be off a little bit, but at least you will know that when you are looking at the weather report in Toronto and the local high temperature is predicted to be 15°, it means you should bring your spring coat, not your wool socks, when you go out. Can you explain why, using a conversion formula? It's because $2(15) + 30 = 60$.

Use a Calculator

Most cell phones have calculators that will do conversions for you. The conversions can be within systems so that (for example) miles can be converted to yards or feet, or between systems so that miles can be converted to kilometers or meters. You do not have to memorize a set of facts to work within the metric system. Having a sense of size will help you when you need to make decisions (about travel time, weight of a box to carry, or the amount of ingredients to use in a recipe).

Arithmetic with Decimals

In terms of adding and subtracting decimal values, when the numbers involved do not have the same number of decimal places, adding zeros to the end of a number does not affect the value of the number. For example, to subtract $25.3 - 16.467$, you can rewrite 25.3 as 25.300 and perform the subtraction mentally or with pencil and paper (if a calculator is not available). Applying the same process you use with the subtraction of whole numbers, you would get

25.300	which	15.300	then	9.300	to	8.900	and	8.840
− 10.000	becomes	− 6.000		− 0.400		− 0.060	finally	− 0.007
15.300		9.300		8.900		8.840		8.833

Be sure to have the decimal points in the same column when adding and subtracting decimals.

The rules of multiplication have a slight trick: The number of decimal places in the product is equal to the sum of the numbers of decimal places in the factors. For example, in multiplying 6.2×0.002, the number of decimal places in the product must be four (one from 6.2 and three from 0.002). Because $62 \times 2 = 124$, the answer to the problem 6.2×0.002 must be 0.0124. This will become a bit clearer when you start multiplying fractions.

Estimate the value of the product: 34.56×6.21

You need an estimate because you want to make sure that you type the decimal points into your calculator in the correct places. 34.56 is approximately 35, and 6.21 is approximately 6. Therefore, $35 \times 6 = (30 + 5) \times 6 = 30 \times 6 + 5 \times 6 = 180 + 30 = 210$. Having this estimate of the product in mind will help you recognize that you need to re-enter the numbers on your calculator if the result you see is 21.46176 (which means you entered 3.456×6.21) or if your result is 2146.176 (oops, you entered 345.6×6.21).

Division with a decimal *divisor* (the number you are dividing by) also has a rule: Just don't divide by a decimal. That is, don't do it if you are going to try the problem by hand. Before going through the rules, it's best to get a mental image of the problem. Take a large candy bar and break it into four equal pieces. Each piece is 0.25 the size of the original bar. How many pieces will you have if you break three bars in the same manner? You might automatically (and rightfully) say 12, because 3 times 4 equals 12. The mathematical problem stated that three candy bars are divided into pieces, each 0.25 the size of a whole bar, so the number of pieces is found with the computation $3 \div 0.25$. This answer must be 12—the logic says so. To do the division problem by hand, create an equivalent problem based on the number of decimal places in the divisor. The divisor 0.25 has two decimal places (if this number is multiplied by 100, it becomes 25). Therefore, the *dividend* (the number you are dividing, in this case 3) is also multiplied by 100, and the new problem is $300 \div 25$. Thus your answer is 12, as expected.

Do you see how much easier it is to follow this calculation when the divisor is a whole number?

Estimate the value: 235.9 ÷ 0.037

First, deal with the decimal points. Since there are three decimal places in the divisor, multiply both the dividend and the divisor by 1,000. The problem is now 235,900 ÷ 37. Rounding 235,900 to 236,000 and 37 to 40, the estimate of the *quotient* (the result in a division problem) is 236,000 ÷ 40, or 23,600 ÷ 4 (cancel a zero from each number, that is, divide 236,000 by 10 and divide 40 by 10) = 5,900.

ALERT

Many students are just told to slide the decimal points in the divisor and the dividend over without any explanation of why this is done. If asked by a student for the rationale of the process, you now can explain it.

Making Change

How many times have you witnessed the following scene? A clerk has rung up an order on the cash register (computer), and the total is, say, $37.13. The customer hands the clerk $40, which the clerk enters into the register, and the display reads that the customer should receive $2.87 in change. At this point, the customer says, "Wait, I have a quarter," and the clerk stares at the screen with a blank look. Stop laughing.

What should the clerk do? The customer has $2.87 due back. The extra 25 cents that the customer handed the clerk is also the customer's money. Therefore, the customer should receive $2.87 + 0.25 = $3.12 back. The customer probably handed over that quarter because 87 cents is heavier to carry than 13 cents—not to mention the weight of the quarter that she or he has off-loaded. (For those who will argue that the clerk should think "The customer gave me $40.25 and the bill is $37.13, then 25 − 13 is 12 and 40 − 37 is 3, so the customer gets $3.12 back," you have a good point. Whether the clerk can do this depends on a few things: Can the clerk see what the original bill is on the register screen and then perform the arithmetic as you suggested? Or, if the register screen does not show the amount of the sale, will the clerk remember it?)

The point is that any extra money the customer gives to the cashier should be added to the change the register records as how much the customer should get back. Parents can use such episodes as "teaching moments" to help their children who are old enough to do such arithmetic understand more about numbers. When they are at the cash register, they can ask their child how much change should be returned if the cashier is given an extra amount (such as that quarter). Money issues and games are the places where children first learn to apply numbers, and the more experience they get, the better off they are.

QUESTION

When do you use decimals?
When you are balancing your checkbook, considering the size of the files you download from the Internet, taking note of the amount of available space on your hard drive, and figuring out the amount of material you will need should you do a craft project. What else did you add to your list of when you use math?

Exercises

Answer the following questions about decimals and their application to everyday life.

1. According to *www.celebritynetworth.com*, Michael Jackson's album *Thriller* is the bestselling album of all time; 65 million albums have been sold. Write this number in scientific notation.
2. Beyond the Milky Way, the galaxy of which we are a part, the Andromeda Galaxy is the closest galaxy to our solar system; it is 2,538,000 light-years from Earth. (A light-year is not a measure of time. It is the distance that light travels in 1 year.) Write this number in scientific notation.
3. Light travels at the speed of 186,000 miles per second. Given that there are 3,600 seconds in an hour, 24 hours in a day, and 365 days in a year,

estimate the distance that light travels in 1 year. Write your answer in expanded notation and in scientific notation.

4. Use your answers from questions 2 and 3 to determine how many miles the Andromeda Galaxy is from Earth. Write your answer in expanded notation and in scientific notation.

5. The temperature in Mexico City is reported as 32°C. Estimate the temperature in degrees Fahrenheit.

6. Kate reported that the temperature in Pembroke, Ontario, during February 2013 was −40°C. Find the temperature in degrees Fahrenheit.

7. The same week that Kate reported that the temperature in Pembroke was −40°C, Diane reported that the temperature in Siesta Key, Florida, reached 85°F. Estimate the Florida temperature in degrees Celsius.

8. Estimate the product: 48.47 × 0.0047

9. Estimate the quotient: 48.47 ÷ 0.0047

10. After giving a cashier $80 for a bill that totaled $67.61, the customer gave the cashier an additional $3. How much money should the cashier return to the customer? Why might the customer have given the cashier the extra $3?

CHAPTER 4

Fractions

There are very few situations in an adult's life where arithmetic with fractions is used. More often than not, people work with decimal equivalents. There are times, such as when hanging curtains and making certain that they are centered on a window, or when centering paintings or posters on a wall, that some addition and subtraction of fractions may occur, but these occasions are rare. The primary purpose of this chapter is to act as a refresher for you in case you should need to explain the mechanics of computing with fractions to someone else.

Fractions Are Numbers, Too

Imagine that you have a stack of one-dollar bills, and you want to group them into packets of 25. You pick up the stack and start counting. You do not say, "1 one-dollar bill, 2 one-dollar bills, 3 one-dollar bills." You count 1, 2, 3, etc., until you reach 25. However, if there is a $5 in the middle of the stack, you take it out and put it aside. Even though the five-dollar bill has the same value as 5 one-dollar bills, it does not serve your purpose as you put together packets of 25 one-dollar bills. Fractions need to be treated in essentially the same way. So long as the common characteristic is the same, you tend not to say the characteristic—you just count. With fractions, the characteristic is the denominator (the number at the bottom of the fraction), and you count the numerators (the numbers at the top of the fraction).

ESSENTIAL

Fractions have a reputation for being difficult to understand because they are usually presented as a set of rules to follow rather than presented in a context. It is no wonder that so many students struggle through the experience and forever have an uneasy feeling when confronted with fractions.

You probably remember your teacher discussing slices of pizza as an illustration of using fractions. If your pizza is cut into 8 pieces, supposedly equal in size, then each slice represents ⅛ of the pizza. You can help someone who is new to learning fractions by referring to the part of the pizza consumed. For example, if John ate 2 slices of the pizza, you could say aloud that John ate ⅖ of the pizza. If the same pizza is cut into 24 slices and Aleks ate 5 slices, then she ate 5/24 of the pizza. If a pizza is cut into 24 slices, how many of these slices represent the same amount of pizza as 1 slice of a pizza cut into 8 slices?

It is important for you to recognize that the two pizzas had to be the same size to begin with in order for you to answer the question in the last paragraph. What if the pizzas were not the same size? You would have to do a lot more work to determine how many slices of each pizza result in an equal amount of food (dough, sauce, and cheese). However, no matter what the size of the pizza, you can talk about the relative amount of the pizza eaten. In this example, eating 1 slice of the 8-cut pizza is the same as eating 3 slices of the 24-cut pizza. Eating 1 slice of an 8-cut pizza means that $\frac{1}{8}$ of the pizza has been eaten. Eating 3 slices of a 24-cut pizza means that $\frac{3}{24}$, or $\frac{1}{8}$, of the pizza has been consumed. That is why $\frac{1}{8}$ and $\frac{3}{24}$ are called *equivalent fractions.*

Adding and Subtracting Fractions

Addition and subtraction involve combining common characteristics. What do you get when you add 8 apples and 3 apples? You get 11 apples. What is the result of taking 4 oranges from 9 oranges? The result is 5 oranges. You can find the sum of 8 and 3 or the difference between 9 and 4 in the abstract, but you can't add 8 apples and 4 oranges and say you have 12 apples, just as you can't say you have 12 oranges.

ESSENTIAL

Unlike the situation where you can look at a bowl of fruit consisting of 5 apples and 4 oranges and know that there are 9 total pieces of fruit in the bowl, you cannot look at the sum (or difference) of two fractions with different denominators and make a claim about them. However, if the fruit bowl contained 5 McIntosh apples and 4 Red Delicious apples, you could claim that there are 9 apples in the bowl—apples being the common denominator of the two entities. The rationale for getting common denominators when adding and subtracting fractions is exactly that—getting the entities to have a common unit of measure (the sixth, the eighth, the twelfth, etc.).

Arithmetic (and mathematics) becomes more interesting when used in context. Going back to the example of the 24-cut pizza, if John ate $\frac{1}{3}$ of the pizza and Aleks ate $\frac{1}{8}$ of the pizza, how much of the pizza remains? In this context, you can calculate how many slices were eaten and how many are left and, from this, determine what fraction of the pizza remains. John ate $\frac{1}{3}$ of the 24 slices. How many slices is this? Well, 24 slices divided by 3 is 8 slices. Aleks ate 3 slices (24 divided by 8). Together, they ate 11 slices. This leaves 13 slices, so $\frac{13}{24}$ of the pizza remains.

From a fractional perspective, John and Aleks ate $\frac{1}{3} + \frac{1}{8}$ of the pizza. Converting these fractions to equivalent fractions with a denominator of 24, we see that they ate $\frac{8}{24} + \frac{3}{24}$ of the pizza. You can consider this to be an example of adding $8 + 3$ with the common characteristic being twenty-fourths, rather than apples or oranges.

$$\frac{8}{24} + \frac{3}{24} = \frac{11}{24}$$

How much of the pizza is left? Aleks and John started with 1 pizza and ate $^{11}/_{24}$ of it. The amount left is $1 - \frac{11}{24}$. Rewriting this with equivalent fractions, $\frac{24}{24} - \frac{11}{24}$, you see that $^{13}/_{24}$ of the pizza is left.

FACT

The prime factorization of a number means you should write the number as a product of the prime numbers that divide evenly into the number. Sounds more complicated than it is. For example, $72 = 8 \times 9$. $8 = 23$ and $9 = 32$. The prime factorization of 72 is 23×32.

Finding a Common Denominator

You recall the mechanics of finding equivalent fractions—multiply both the numerator and the denominator by the same value. For example, add: $\frac{5}{6} + \frac{7}{8}$

We need to find the "special numbers" that can be used to create the common denominators with as small a set of numbers as is possible. You may immediately recognize the greatest common factor of 6 and 8. If not, rewrite both denominators with their prime factors: $6 = 2 \times 3$; $8 = 2 \times 2 \times 2 = 2^3$. You can see that the largest factor that both numbers have is a 2 in common. Therefore the greatest common factor of 6 and 8 is 2.

Therefore, $\dfrac{5}{6}+\dfrac{7}{8}=\dfrac{450}{2900}=\dfrac{n}{100}\ \dfrac{5}{3\times2}+\dfrac{7}{4\times2}$. If you multiply the first fraction by ¾ and the second fraction by ⅔, you get

$$\frac{5}{6}+\frac{7}{8}=\frac{5}{3\times2}+\frac{7}{4\times2}=\frac{5}{3\times2}\ \frac{4}{4}+\frac{7}{4\times2}\ \frac{3}{3}=\frac{5\times4+7\times3}{4\times3\times2}=\frac{41}{24}$$

Note that $\dfrac{41}{24}$ is an *improper fraction* (one in which the numerator is larger than the denominator). It is common to write such a fraction as a mixed number, in this case $1\dfrac{17}{24}$. You can find this number by dividing 41 by 24 and writing the remainder, 17, over 24 as the proper fraction.

FACT

> A proper fraction has a numerator that is smaller than its denominator; an improper fraction has a numerator that is greater than or equal to the denominator. A mixed number contains a whole number and a proper fraction.

Here is a more involved problem that illustrates the addition process:

Add: $\dfrac{23}{168}+\dfrac{37}{210}$

$168=2\times84=2\times2\times42=2\times2\times2\times21=2\times2\times2\times3\times7$

$210=2\times105=2\times5\times21=2\times5\times3\times7$
168 and 210 have $2\times3\times7=42$ as common factors.

$$\frac{23}{168}+\frac{37}{210}=\frac{23}{42\times4}+\frac{37}{42\times5}$$

Get equivalent fractions:

$$=\frac{23}{42\times4}\ \frac{5}{5}+\frac{37}{42\times5}\ \frac{4}{4}=\frac{115}{840}+\frac{263}{840}$$

The size of the denominators does not matter. Learning how to find a common denominator is the key to adding and subtracting fractions.

Calculate: $4\dfrac{5}{8} - 2\dfrac{7}{8}$

The denominators are the same, but there is a need to borrow. The traditional approach is to borrow 1 from the 4, rewrite the $1\dfrac{5}{8}$ as $\dfrac{13}{8}$, and then proceed to subtract $3\dfrac{13}{8} - 2\dfrac{7}{8}$ to get $1\dfrac{6}{8} = 1\dfrac{3}{4}$.

You could also rewrite each fraction as an improper fraction and then subtract. In this approach, $4\dfrac{5}{8}$ becomes $\dfrac{37}{8}$ (multiply the denominator by the whole number and add the numerator), and $2\dfrac{7}{8}$ becomes $\dfrac{23}{8}$. Subtract $\dfrac{23}{8}$ from $\dfrac{37}{8}$ to get $\dfrac{14}{8} = 1\dfrac{6}{8} = 1\dfrac{3}{4}$.

Multiplying Fractions

If you take a pizza and divide it into 8 (equal) slices, each slice represents ⅛ of the pizza.

$$1 \div 8 = \dfrac{1}{8}$$

If you take one of these slices and divide it into 2 equal pieces, what part of the original pizza is each of these new slices? Cutting each of the original slices in two would give 16 slices. Therefore, the answer to the last question is that each smaller piece will be ⅟₁₆ of the pizza. Thus $\dfrac{1}{8} \div 2 = \dfrac{1}{16}$.

You can also look at this problem as taking ½ of each slice (⅛ of the pizza). Thus $\dfrac{1}{8} \times \dfrac{1}{2} = \dfrac{1}{16}$.

In each case, the ⅛ represents how much pizza, and the ÷ 2 or × ½ represents the action performed on the pizza. For this reason, a common denominator is *not* needed.

FACT

In addition, note that multiplying by ½ is equivalent to dividing by 2.

Multiply: $\dfrac{15}{64} \times \dfrac{8}{25}$

The basic rules for multiplication are that numerators are multiplied together and denominators are multiplied together. A consequence of this is that the numbers can get very large. To get a "more reasonable, reduced" answer, factor the numerator and denominator.

$$\frac{15}{64} \times \frac{8}{25} = \frac{3 \times 5}{8 \times 8} \times \frac{8}{5 \times 5} = \frac{3 \times 5 \times 8}{8 \times 8 \times 5 \times 5} = \frac{3}{8 \times 5} \times \frac{5 \times 8}{5 \times 8} = \frac{3}{40}$$

Yes, you remember a quicker way to reduce this problem. But before you do that, be sure you understand that what you are about to do is really the same process as that shown earlier, without the need to put all the factors into one fraction; the only operation in the numerator and in the denominator is multiplication, and "canceling" is really just dividing the numerator and denominator by the same number. You will find that you won't struggle with this in multiplication and division problems as you might when facing an addition or subtraction problem.

ESSENTIAL

A man-hour is a term used to describe the amount of labor used to complete a job (for example, if 5 men are hired for 8 hours each, the company will charge for 40 man-hours of labor). There is no common characteristic between the number of men working and the number of hours each works. Yet, they can be multiplied together. Similarly, the fractions $\dfrac{2}{3}$ and $\dfrac{5}{8}$ can be multiplied even though they do not have a common characteristic.

Dividing Fractions

Let's consider another pizza problem. Suppose you have 12 pizzas, and you are trying to decide how many people to invite to a party based on the amount of pizza each person will eat. Consider the following table:

PIZZA PARTY

Amount of Pizza per Person	Number of People
2 pizzas	
1 pizza	
½ pizza	
⅓ pizza	
⅛ pizza	
⅔ pizza	

How many people can you invite?

- 12 pizzas / 2 pizzas per person = 6 people
- 12 pizzas / 1 pizza per person = 12 people
- 12 pizzas / (½) pizza per person = 24 people

Let's talk about the fraction ½. This represents 1 pizza for 2 people. So you would multiply the number of pizzas by 2 to determine the number of people.

- 12 pizzas / (⅓) pizza per person = 36 people

Providing 1 pizza for every 3 people means that you need to multiply the number of pizzas by 3 to determine the number of people.

- 12 pizzas / (⅛) pizza per person = 96 people

Allotting 1 pizza for every 8 people means that you need to multiply the number of pizzas by 8 to determine the number of people.

- 12 pizzas / (⅔) pizza per person = 18 people

QUESTION

If a snail crawls at a rate of $\frac{5}{8}$ of a foot per hour, how long would it take the snail to crawl 3 feet?

$3\,feet \sqrt{\dfrac{5\,feet}{8\,hours}}$ becomes $3\,feet \times \dfrac{8\,hours}{5\,feet}$ and this equals $\dfrac{24}{5}$ hours or 4.8 hours.

The ⅔ is interpreted to mean that you will need 2 pizzas for every 3 people. So you need to divide the number of pizzas by 2 and multiply this result by 3.

This illustration should give you some insight into the rule for dividing fractions: To divide fractions, change division to multiplication and take the reciprocal of the divisor. How many people should be invited if each person will eat ¾ of a pizza? How many should be invited if each will eat ⅙ of a pizza?

ALERT

It is often the case that the units in a problem are ignored when performing a calculation. For example, if driving a distance of 420 miles at a speed of 60 mph (miles per hour), most people can tell you that it will take 7 hours to make the trip. A closer look at the units shows that $miles \sqrt{\dfrac{miles}{hour}}$ is not as clear as $miles \times \dfrac{hours}{mile}$. The units of miles cancel leaving only hours.

Converting Between Fractions and Decimals

Converting a fraction to a decimal is done easily with a calculator. The way fractions are written (for example, $\frac{5}{8}$) indicates that 5 should be divided by 8 to get the result 0.625. Most people are comfortable with the fact that $\frac{1}{2}$ is equal to 0.5, $\frac{1}{4}$ is equal to 0.25, and $\frac{3}{4}$ is equal to 0.75. Many people know that the fractions $\frac{1}{3}$ and $\frac{2}{3}$ are equal to the repeating decimals 0.333 . . . and 0.666 . . . , respectively. With practice, you can work with a few more of these equivalences by applying simple arithmetic. Half of $\frac{1}{2}$ is $\frac{1}{4}$, so half of 0.5 is 0.25. In the same way, half of $\frac{1}{4}$ is $\frac{1}{8}$ and half of 0.25 is 0.125, so $\frac{1}{8} = 0.125$. You could extend this to half of $\frac{1}{8}$ is $\frac{1}{16}$, so $\frac{1}{16} = 0.0625$. More important, you can now say that because $\frac{1}{4} + \frac{1}{8} = \frac{3}{8}$, $\frac{3}{8}$ must be equal to 0.25 + 0.125 = 0.375.

QUESTION

What is the decimal equivalent of $\frac{7}{8}$?
$\frac{7}{8} = \frac{3}{4} + \frac{1}{8}$, so $\frac{7}{8} = 0.75 + 0.125 = 0.875$

The initial step for converting decimals to fractions is also straightforward. The trick is to read the decimal without using the word *point*. It is customary to read the decimal 0.4 as "point 4." But the number 0.4 is really "4 tenths." The number 0.4, when written as a fraction, is $\frac{4}{10}$. In reduced form, $\frac{4}{10}$ is equal to $\frac{2}{5}$.

Rewriting 0.4 as a fraction is an easy example. Writing 0.076 as a fraction in reduced form takes a little more work. First, read the decimal as 76 thousandths. Write the fraction as you just read it, $\frac{76}{1000}$. Use the divisibility tests (see Chapter 2) to recognize that 76 and 1,000 are both divisible by 4, so $\frac{76}{1000} = \frac{19}{250}$.

Use a Device

The four basic functions on a calculator include division, so more complicated fractions are easily converted to decimals. These calculators cannot convert decimals to fractions, however. The software program Excel does convert some decimals to fractions. Within the menu for Format, choose Cells. The window with the formatting option appears, and the first tab is Number. One of the options within this menu is Fraction. There are limitations on the decimals the program can convert (for instance, there can be no more than three digits in the denominator), but even with these limitations, Excel should meet the needs of most people.

ALERT

Metric sizes corresponding to wrenches measured in sixteenths (¹⁄₁₆ of an inch) are either 1 mm or 2 mm away from the next smaller wrench size measured in eighths (¹⁄₈ of an inch). It is more than likely that you will be looking for the wrench in standard measurement that corresponds to a 21-mm metric wrench. Although 20.5 (not 21) is midway between 19 and 22 (which correspond to the $\frac{3}{4}$- and $\frac{7}{8}$-inch wrenches, respectively), the size you want is the $\frac{13}{16}$-inch wrench.

Hand Me the Wrench

Have you ever purchased an item that needed to be assembled and found that the hardware was produced in a country that uses metric sizes rather than the standard measurement you are accustomed to? Frustrating, right?

The conversions between wrenches in standard measurement and those in metric measurement (available with a search on an Internet browser) work in the following manner: A 1-inch wrench corresponds to a 25-mm wrench; a $\frac{1}{2}$-inch wrench corresponds to a 13-mm wrench. Did you note that half of 25.4 is 12.52, so the number representing the size is rounded up? A $\frac{3}{4}$-inch wrench (midway in size between a $\frac{1}{2}$-inch wrench and a 1-inch wrench) corresponds to a 19-mm wrench (midway between 13 mm and 25 mm).

What size metric wrench corresponds to a $\frac{5}{8}$-inch wrench?

Because $\frac{5}{8}$ is midway between $\frac{1}{2}$ and $\frac{3}{4}$, the corresponding metric wrench should be midway between 13 mm and 19 mm; that is, it should be a 16-mm wrench.

QUESTION

When do you use fractions?
Consider when you are cooking or using a ruler (or tape measure) to measure a piece of wood or cloth before cutting it. What did you add to your list of when you use math? What else did you add to your list of when you use math?

Exercises

Answer these questions about fractions.

1. Which of these fractions are equivalent to $\frac{3}{4}$?

    (a) $\frac{30}{40}$ (b) $\frac{9}{15}$ (c) $\frac{12}{16}$ (d) $\frac{45}{60}$

2. Find the sum of $\frac{5}{6}$ and $\frac{5}{8}$.

3. Find the sum of $\frac{9}{56}$ and $\frac{5}{63}$.

4. Calculate: $\frac{7}{12} - \frac{5}{36}$

5. Calculate: $12\frac{7}{12} - 7\frac{5}{8}$

6. Calculate: $\frac{25}{36} \times \frac{21}{40}$

7. Calculate: $5\dfrac{5}{6} \times 6\dfrac{1}{4}$

8. Calculate: $5\dfrac{5}{6} \div 6\dfrac{1}{4}$

9. Convert $\dfrac{9}{16}$ to a decimal.

10. Convert 0.45 to a fraction.

11. What size wrench in standard measurement corresponds to a wrench with metric size 32 mm?

CHAPTER 5

Percentages

One of the most important practical math skills you can have is a good understanding of percent. Percent is used in every aspect of a person's life, including everything from tipping to salary raises (or cutbacks), discounts, sales taxes, income tax, automobile interest, and mortgage rates. In this chapter, you will master conversions involving percent, and you will examine some of the applications of percent that can be performed either mentally or with a simple calculator (such as you might find on a cell phone, on your computer, or around the house).

What Is Percent?

What does percent mean? *Percent* literally means "out of 100." The first thing you want to do when dealing with percentages is to translate them to decimals (or convenient fractions). Converting a percentage to a decimal is easily accomplished by sliding the decimal point of the number two places to the left. For example, 10% becomes 0.10; 37% becomes 0.37; 6.5% becomes 0.065. Although it is not absolutely necessary to write the 0 before the decimal point, doing so makes the location of the decimal point clearer. The number 50%, or 0.50, is also easily written as ½ (because 50/100 is the same as ½ when reduced). Four other converted percent values that are useful to know are 25% (¼), 75% (¾), 33 ⅓% (⅓), and 66 ⅔% (⅔).

> Example: Convert 24% to a decimal.
> Answer: Drop the percent symbol and move the decimal point two places to the left, to get 0.24.
> Example: Convert 0.045 to a percent.
> Answer: Move the decimal point two places to the right and include the percent symbol, to get 4.5%.

FACT

To change a percent to a fraction, remove the percent sign and write the number as the numerator and the denominator as 100. Reduce the fraction if possible. For example, 52% becomes $\frac{52}{100}$ and that reduces to $\frac{13}{25}$.

Now let's investigate further by converting 25% to a fraction. Remember, *percent* means "out of 100," so 25% is 25 out of 100, or $\frac{25}{100}$.

But now you need to simplify. What is the biggest number that will divide evenly into both 25 and 100? Divide the top and bottom numbers of the fraction by that number like this:

$$\frac{25 \div 25}{100 \div 25} = \frac{1}{4}$$

Thus 25% is the same thing as $\frac{1}{4}$.

Now, convert 25% to a decimal: When you convert percents to decimals, you're actually dividing the percent by 100. $25\% = 25 \div 100 = 0.25$. But there's a pattern here that makes things a lot easier. All you are doing is moving the decimal point two places to the left. (Check it with your calculator, if you're suspicious.) $25\% = 0.25$. There's no decimal point in 25%, you say? Actually, there is. All numbers have decimal points; if you don't see them, it's just because they've been dropped, and you can put them in at the far right of the number. So the decimal point in 25% is on the right side of the 5, for 25.00. Move it two spaces to the left, to produce 0.25.

Calculating a Percent with a Proportion

Before tackling percentages, it is important to review how to solve a proportion. After writing the proportion, cross-multiply to get an equation that is easily solved.

ESSENTIAL

Proportions are solved by a process known as cross multiplication. If the proportion is $\dfrac{a}{b} = \dfrac{c}{d}$, the result after cross-multiplying is $ad = bc$.

Example: Solve: $\dfrac{n}{18} = \dfrac{20}{45}$

Cross-multiply (that is, multiply the top left with the bottom right; multiply the top right with the bottom left), and set these two products equal to each other, to get $45 \times n = 18 \times 20$, or $45 \times n = 360$. Then divide both sides of the equation by 45 to get $n = 8$.

How would you use this skill? In the context of percentages, a simple problem might look like this: Find 30% of 180. Another way to say this is "Find out what number is the same part of 180 as 30 is of 100."

To solve this problem, first write the proportion $\dfrac{n}{180} = \dfrac{30}{100}$. Cross-multiply to get $100 \times n = 180 \times 30$, or $100 \times n = 5400$. Divide by 100 to get $n = 54$.

Example: 84 is 5% of what number?

Until you get more comfortable with the variations in the way these problems can be written, you will want to write (or at least mentally restate) the problem in the form "<blank>% of <some number> is <some other number>." For this example, you will write 5% of what number is 84? Translating this sentence into a proportion gives $\dfrac{84}{n} = \dfrac{5}{100}$. (Note that this second fraction literally means *5 per hundred*, or *5%*.) Cross-multiply to get $5 \times n = 84 \times 100$, or $5 \times n = 8400$, and divide by 5 to get $n = 1680$.

The last variation on this problem is to find a percent. For example, what percent of 2900 is 450?

Write the proportion $\dfrac{450}{2900} = \dfrac{n}{100}$. Multiply to get $2900 \times n = 450 \times 100$, or $2900 \times n = 45{,}000$. Divide by 2900 to find that n is approximately 15.517. Depending on the nature of the problem and the degree of accuracy required, you might say the answer is 15.5% or 15.52%

ESSENTIAL

Percents are often used to describe a relative change. For example, if the price of a gallon of gas rises from $3.25 per gallon to $3.50 per gallon (a 25-cent increase), the percent change is 7.7%. If the price of gas goes from $3.50 per gallon to $3.75 per gallon (again, a 25-cent increase), the percent change is 7.1%.

Calculating a Percent from a Formula

Although at first it may seem less intuitive, a more practical translation of the percent relationship is "*r*% of a base is the part." (Here the letter *r* is substituted for the more cumbersome "blank" that was used in the earlier discussion.) The equation for this sentence, then, is

$$r\% \times base = part$$

This equation is much easier to apply than the first method of using proportions (whether with mental math or with a calculator). The only requirement is that the term *r*% be rewritten as a decimal or a fraction (although there are some calculators that will do this step for you).

Example: Find 20% of 80.
Answer: $0.20 \times 80 = 16$
Example: Find 6.8% of $12,000.
Answer: $0.068 \times \$12000 = \$816$

The percent can be found by dividing the part by the base.

Example: What percent of $900 is $150?

Answer: Because the rate (percent) is unknown, the equation reads $r \times 900 = 150$. Divide by 900 to get the decimal equivalent of the percent, $^{150}\!/_{900} = 0.166666666667$. Next convert this to a percent by moving the decimal point two places to the right, to get 16.6666666667%, which, when rounded to one decimal place, is 16.7%.

You may be thinking that you could just do this on a calculator, and you're right! But keep in mind that not all calculators are created equal, especially with regard to calculating percents. To multiply 200 by 5%, you might type $200 \times 5\%$, some calculators will display the answer 10 while other calculators will display 0.05 and wait for you to press the equals key. If you are using a calculator for these exercises, be sure you are familiar with how your calculator works in this context.

Finding the Percent Change

A retail store announces that the prices on all its items are being reduced by at least 10%. The announcement also states that the price of a television is reduced from $1,800 to $1,635. Is this price change consistent with the announcement that all prices will decrease by 10%? To calculate the percent change in price, you must first compute the difference and determine the percent of the original price that this number represents. So you begin by subtracting $1,800 − $1,635 to get $165. Then you find the percent:

$r \times 1800 = 165$
$r = 165 \div 1800$, which is about equal to 0.0917, or 9.17%

The price of the television has been reduced by 9.17%, and this is *not* consistent with the store's claim that all prices are reduced by at least 10%. Unfortunately (for the customers, that is), a number of people incorrectly assume that because 10% of the *sale price* $1,635 is $163.50, a price reduction of $165 supports the store's claim. But the math isn't correct. The process for finding percent change must always be based on the *original value*. Clearly, when you are assessing the value of a reduced-price purchase, the best advice to follow is "You do the math!" When you do, you will find that to honor the claim made in its advertisement, this store should have priced the television in question no higher than $1,620 (that is, $1,800 − $180).

Here is another example of the need to base percent change on the original value: The sports network ESPN reported that in December 2012, the New York Yankees paid a luxury tax (an amount paid for awarding salaries that exceed the salary cap established by Major League Baseball) of $18.9 million, up from $13.9 million the year before. What was the percent change in the luxury tax that the Yankees paid from the 2011 baseball season to the 2012 season?

The change between the two seasons was $5 million, so the question is "$5 million is what percent of $13.9 million?"

$r \times 13.9 = 5$

$r = 5 \div 13.9$, which is about equal to 0.3597

The percent change from 2011 to 2012 was 35.97%.

Increasing and Decreasing Rates

Reports about changing rates are common in the news media. The example of the New York Yankees luxury tax, in the previous section, is just one instance. But many people don't understand exactly what they mean. When you encounter such numbers in the news, make a habit of mentally rehearsing the supporting calculations and paraphrasing their meaning, as if you were explaining it to someone who didn't understand. No calculator or pencil needed—just an alert awareness.

ESSENTIAL

When calculating a rate of increase or decrease, it is always the amount of change divided by the value before the change. When the price of gasoline changed from $3.50 to $3.75, the change is $0.25 and the percent increase is $\frac{0.25}{3.50} \times 100$, or 7.7%. Had the price dropped from $3.75 to $3.50, the percent decrease would be $\frac{0.25}{3.75} \times 100$ or 6.7%.

On March 26, 2013, *NBC Nightly News* reported that there were 255 houses for sale in Alexandria, Virginia, and that this represented a 31% drop from the number of houses that were for sale in the previous year. Many people hearing this would want to determine the number of houses on the market in Alexandria the year before. Their analysis might sound like this: "I'll use 30% as an estimate of the rate. I know that 10% of 255 is 25.5, so 30% must be 76.5. Because there were roughly 76 more houses on the market last year, the total for last year is about 331." The trouble with this analysis is that the 30% change in available homes should have been based on the original value—that is, on the number of homes available in 2012, a number larger than the 255 homes available at the time of the report.

Context

The correct approach to answering this question is to put the values in the correct context. The number of houses on the market at the time of the report is 31% fewer than the number of houses available in 2012. This means that the number of houses available in 2013 is 100% − 31% = 69% of the houses available the previous year. The new question is "255 is 69% of what number?"

$$225 = 0.69 \times n$$
$$n = 225 \div 0.69 = 370$$

The answer is 370, a difference of 39 houses.

The downturn in the economy in recent years has caused a number of businesses to lay off workers. To illustrate the importance of the base number in any discussion of rates of change, consider the following hypothetical situation. Say the owners of a factory had to lay off 25% of their employees as a consequence of poor economic conditions. Some time later, when the owners received a new contract, they rehired a number of those who had been laid off equal to 25% of the current work force. Is the factory back to full employment? Well, suppose the owners originally had 400 people working for them. They laid off 25%, or 100, of these workers, when the economy was poor, leaving them with a work force of 300. Then, after a time, they rehired a number of those who had been laid off equal to 25% of that work force of 300. This means they rehired 75 people, bringing their total work force to 375, *not* back to the 400 they originally employed.

QUESTION

Where do you use percents in your life?
Consider when you shop, when you read the newspaper or watch the news and hear the economics reports relating the current market to a time in the past, or when you receive a pay raise. What else did you add to your list of when you use math?

Exercises

Answer the following questions about percent.

1. Change 80% to a fraction in lowest terms.
2. Find 80% of 120.

3. Find 17.2% of 15,600.
4. What percent of 28 is 4.6?
5. What percent of 342 is 456?
6. 78 is what percent of 90?
7. 78 is 8% of what number?
8. The price of gasoline jumped from $3.29 to $3.72 in a single month in Florida. What was the percent increase in the price of gasoline during that month?
9. On March 28, 2013, *Newsday* reported on its website (*http://newyork.newsday.com*) that *Forbes* magazine estimated that the average Major League Baseball (MLB) team is worth $744 million, an increase of 23% since 2012. What was the average value of an MLB team in 2012?

CHAPTER 6

Shopping

The place where you probably use mathematics the most is at the store or in a restaurant. When you are shopping, not only do you end up buying items that are on sale or that you have a coupon for, but often you also have to pay a sales tax. Comparing prices is an essential skill for any shopper. Learning how to do quick math to get yourself the best deal can save you time and money. You may also need to figure out a tip when you stop for lunch during a shopping trip. The wait staffs in American restaurants are paid an hourly wage that is very close to the minimum wage because the expectation is that they will earn tips from customers rewarding them for the service they provide. Like estimating sales taxes, computing tips is an important skill that you will use regularly.

How Much Did You Save?

The sign at the entrance to your favorite store reads, "25% Off All Items in the Store." You enter the store and find a coat that you've been wanting but did not buy because, at a price of $240, the coat was too expensive. How much money will you save if you purchase the coat at the sale price? Knowing that 25% is the same as the fraction $\frac{1}{4}$, you multiply the fraction by $240 and find that you will save $60 on the cost of the coat. You already know the coat fits perfectly. Now you just need to decide whether the sale price of $180 ($240 − $60) fits in your budget.

ALERT

Using coupons when shopping can save you a significant amount of money whether the discount is in terms of a flat dollar fee or a percent of the tag price.

While knowing the amount of the discount is important (and maybe a bit exciting), knowing the final price of an item being sold at a discount is more important. In this case, you need to ask yourself the question, "What percent of the original price am I paying?"

Example: A coat is being sold at a 35% discount. If the original price for the coat is $210, what price will Stephanie pay after the discount?

During this sale, the buyer of the coat will pay 65% (100% minus 35%) of the original price. Use a calculator to determine that the sale price is $0.65 \times \$210 = \$136.50$. (The buyer saves $73.50 on the purchase.)

How Much More Did You Save?

Many stores offer shoppers who frequent their store a coupon that entitles them to an additional discount. It is important to note that such a coupon does not increase the original sale discount; rather, it acts as a separate discount. For example, if the store is having a 60% sale and the coupon entitles the shopper to another 10% discount, the percent of the original price paid by the shopper is $0.40 \times 0.90 = 0.36$ (100% − 60% × 100% − 10%), or 36% of the original price rather than 30% (100% − 70%, the sum of the 60% and 10% discounts).

ESSENTIAL

Rather than computing the amount you are saving from the discount, you can use the percent you are still paying to compute the sale price. For example, a $150 dress sold at 40% off requires that you pay 60% of the $150. 60% of $150 is $90.

For example, say Dave and Angie Netoski want to buy a sofa with a retail price of $1,280 that is advertised for sale at a 30% discount. The Netoskis are interested in buying the sofa, especially since Angie has a coupon to this store giving her an additional 15% discount. To their great surprise, when they enter the store, Dave and Angie notice a sign saying that the manager is having an additional 5% discount taken at the register for any purchase in excess of $500. What price will the Netoskis pay for the sofa?

ALERT

When computing successive discounts, multiply each of the percents that you must pay by the *original price* to determine the final sale price.

Angie and Dave will pay 70% of the retail price based on the store's sale. Because of Angie's coupon, they will pay 85% of the sale price, and then they will pay 95% of that figure because of the manager's special. The price they will pay is $1,280 × 70% × 85% × 95% = $723.52. Note that this calculation is expressed in terms of what percent the customer will *pay* at each stage of applying successive discounts. Another approach is to express the situation in terms of what the customer will *save* at each step. Speaking in those terms, we would say that when more than one discount is applied to a purchase, the second discount is applied to the price that resulted when the first discount was taken, and any third discount is applied to the price that resulted when both earlier discounts had been taken. The key thing to remember is that getting a 30% discount, a 15% discount, and a 5% discount on an item does *not* mean you will get a 50% discount off the original price, as you might conclude if you simply added the three discounts. Instead, each discount is applied one at a time, after taking the previous discount(s). (Otherwise, our friends Dave and Angie would have paid only $640 for the thrice-discounted sofa that cost them $723.52.) There is nothing illegal in the way the store advertised the discounts it offered, but if you want to accurately estimate the final cost of items where multiple discounts are applied, you need to remember that the discounts are taken successively.

Sales Tax

Being able to compute the final sale price you will pay is more important than knowing how much money you saved on the purchase price (because you need to be sure you have enough money to pay for the purchase). Similarly, knowing the cost after tax is more important than knowing the price before the addition of any sales tax(es). After all, your goal is to find the final amount that you will need to pay.

FACT

The final price of a purchase is 100% of the cost of the objects you buy plus the tax that must be paid. In an area where the sales tax is 6.5%, the final cost of the purchase will be 106.5% of the cost of the objects purchased.

For example, Stephanie is buying a coat originally priced at $210 for the sale price of $136.50 during a 35%-off sale. What is the final price of the coat if the sales tax is 8%?

Well, the final price is 108% of the sales price: $1.08 \times \$136.50 = \$147.42$.

The calculation of the price after tax can actually be done in one step. Starting with the original price, and noting that during a 35%-off sale she will pay 65% of that original price, Stephanie could have computed the sale price, and then the total price with tax, simply by multiplying the three numbers $\$210 \times 0.65 \times 1.08 = \$147.42$.

Here's another example to consider: A department store offers a 60%-off sale for holiday decorations once the holiday has passed. The sales tax in the region is 4%. What is the total price, tax included, for the purchase of holiday decorations that had an original price of $45.58?

Answer: With a 60% discount, the buyer is paying 40% of the original price. The final price with tax will be 104% of the sale price. $45.58 × 0.40 × 1.04 = $18.96128, which the store will round up to $18.97.

Example: On the same day, a shopper at the same department store purchases $83.47 worth of decorations. When he goes to check out, he gives the cashier a coupon that entitles him to an additional 10% discount. He must also pay a sales tax of 4%. Determine the final price he paid for the decorations.

Answer: The final price is $83.47 × 0.40 × 0.90 × 1.04 = $31.251168, which the store will round up to $31.26.

Unit Pricing

The purpose of unit pricing is to provide the consumer with information to use in deciding which product to purchase. When you go to the delicatessen section of the supermarket, you can easily see the price of the meat you are buying. The store-brand oven-roasted turkey sells for $6.99 per pound, while a national brand sells for $7.99 per pound. In each case, you know the price you are paying for each pound of the product. That is, you can see the unit price for each item. Although you can (and should!) take the unit price into account in deciding which product to buy, you also need to consider the quality of the product you are getting. When there is a significant difference in taste, for example, there is no sense in purchasing the cheaper product if you will not eat it.

ESSENTIAL

Even when the unit prices are on display, be sure to read the units involved when comparing prices. A product that posts its unit cost per pound will appear to be more expensive than a product that posts its unit cost per ounce.

Do the Math

Say a gallon of skim milk costs $3.59, and a half-gallon of the same brand of skim milk costs $2.22. What is the price per quart for each container?

First you need to do some conversion. There are 4 quarts in a gallon, so the gallon of milk costs about $0.90 ($3.59 ÷ 4) per quart, and the half-gallon container costs $1.11 ($2.22 ÷ 2) per quart.

Example: A 5-ounce can of StarKist tuna sells for $1.59. What is the price per pound?

First you divide $1.59 by 5 (it's a 5-ounce can), and then you multiply by 16 (because there are 16 ounces in a pound). The price per pound for the tuna is $5.09.

Example: The same store is having a sale on 5-ounce cans of Bumble Bee tuna—4 cans for $5. How much money will you save if you buy 2 cans of the Bumble Bee tuna rather than 2 cans of the StarKist tuna?

Two cans of Bumble Bee tuna cost $2.50, and two cans of StarKist cost $3.18. You save $0.68.

Mistakes can be made, which is why it is always a good idea to have a calculator on hand as you shop. On a recent shopping trip, Diane was comparing two pasta sauces, Ragu and Prego. Each jar contained 24 ounces of sauce. The pricing information for the jar of Prego showed a retail price of $2.79 and a unit price of $1.86 per pound. The information for the jar of Ragu showed a price of $2.29 and a unit price of $1.41 per pound. Diane took out her cell phone, opened the calculator tool, and checked the numbers. The unit price for the Prego was $2.79 ÷ 24 × 16 = $1.86 (as the label indicated), but the unit price for the Ragu should have been $2.29 ÷ 24 × 16 = $1.53. The unit price for the Ragu was incorrect. (In this instance, Ragu is still the cheaper of the two purchases, but sometimes when errors occur, the wrong product is made to appear to be the less costly. In addition, note that many shoppers believe that the biggest package of a product always has the lowest unit price. This is generally true, but the adage "Let the buyer beware" still applies. Always check the unit pricing on products to determine which is the best deal for you.

Eating at a Restaurant—Tipping

Tipping your server after a meal at a restaurant is common in many countries. In fact, in some countries (such as France) the tip is included in the bill. The percent you tip the wait staff depends on individual preference. Exceptionally good service may encourage the consumer to increase the tip, whereas poor service may result in a lower tip. Many cell phones have a Tip feature as part of the calculator tool set. Some restaurants have their registers programmed to recommend tips at different rates (such as 15%, 18%, and 20%) to assist their customers (and to remind them to remember to tip).

The following list offers some easy tips (pardon the pun) for computing the gratuity (tip) for wait staff.

Calculating Tip

No matter where you are in the world, you're expected to tip. There's a simple way to figure out the tip on any bill—whether it's for drinks, a meal, a taxi, or a tour guide. First, round your total. There's no reason to kill brain cells figuring the tip on $37.99, when $38 is so close. If you had really good service, 20% of the bill is standard. To find it, divide the total by 5 (20% is the same thing as $\frac{1}{5}$). Suppose your bill is for $45.60. Round up to $46. Then $46 \div 5 = 9\frac{1}{5}$, or $9.20. Or, find 10% of the total, and double it.

10% of 46 is $4.60
$2 \cdot \$4.60 = \$9.20$

Finding 15% is a little trickier, but it can be done mentally in three little steps. Just find 10% of the total. Then halve the total (that's 5%). Then add these two results.

10% of 46 is $4.60
$\$4.60 \div 2 = \$2.30$
$\$4.60 + \$2.30 = \$6.90$

It is perfectly reasonable to base the tip for service rendered on the pre-tax total of the bill.

Here are a few more examples to help you on your way:

- 15%: To compute a 15% tip, determine the amount of the check, slide the decimal point one place to the left to get 10%, take half of that number, and add. Thus, if the bill were $10.00, you would tip $1.00 + $0.50 = $1.50.
- 20%: To compute a 20% tip, determine the amount of the check, slide the decimal point one place to the left to get 10%, and double that number. Thus, if the bill were $10.00, you would tip $1.00 × 2 = $2.00.

To tip between these two rates, it may be easiest to determine the 15% and 20% values and then decide on a number between these that makes sense. Some people like to tip in whole dollars, some to the nearest quarter, others in such a way that the bill, with tip, adds up to a total ending in a whole dollar. It is all a matter of personal taste.

For example, Gary and Cathy received a bill of $67.78 for their dinner. How much should they leave for a tip if they want to leave a 15% tip? How much should they leave if they want to leave a 20% tip?

Answer: Sliding the decimal point to the left one place gives $6.78. Half of this number is $3.39, so a 15% tip would be $6.78 + $3.39 = $10.17. (In round numbers, $10 might be considered appropriate.) To compute the 20% tip, double $6.78 to get $13.56. ($13.50 might be appropriate.)

Example: What percent tip would Gary and Cathy have left if they gave a tip of $12 for a bill of $67.78?

Answer: Divide $12 by $67.78 to get 0.177, or 17.7%.

QUESTION

When do you use math when you are shopping?

Consider the money you save using coupons, buying items on sale, and BOGO (Buy One, Get One). What else did you add to your list of when you use math?

Exercises

Stephanie and her friend Bear go shopping. Answer the following questions that arise from their shopping experience.

1. Bear sees a dog bed that normally sells for $45.80 and is on sale for 30% off. How much will they save if they buy the dog bed?

2. Stephanie sees a different dog bed that has a retail price of $59.89. This bed is being sold for 40% off, and Stephanie has a coupon from the manufacturer that will entitle her to an additional 10% off. What is the sale price for this bed?

3. They decide to purchase the cheaper of the two dog beds. What is the final cost of the purchase after they pay the 7% sales tax?

4. They decide to have lunch at a local restaurant. The lunch bill totals $29.85 before taxes. Bear has a 15%-off coupon for the restaurant. Given that the sales tax is 7%, find the total cost of the meal.

5. Bear believes that the tip he leaves should be based on the pre-discounted price (but before taxes). Bear tips by paying 20% of the price (and he will round his tip to the nearest dollar). How much will Bear leave as a tip?

6. After lunch, Stephanie and Bear go to the grocery store. When they get to the aisle containing cereals, Stephanie particularly likes Honey Nut Cheerios. There are three choices: a box containing 12.5 ounces at a cost of $3.99, a 17.5-ounce box at a cost of $4.99, and a 21.5-ounce box at a cost of $6.99. Which box offers this cereal at the lowest cost per pound?

7. While driving home from the supermarket, Stephanie was telling Bear about the lamp she bought earlier in the week. Knowing her well enough to be sure that the lamp she purchased was on sale, Bear asked what the original price of the lamp was. "I don't remember," Stephanie responded. "I do know that the sale was for 40% off and of course I paid the 7% sales tax. I know I paid a total of $34.16 for the lamp, including the tax. I really like the way it looks, and it seems like a good price." They had just arrived at home, so Bear whipped out his cell phone, opened the calculator tool, and determined the original price of the lamp. What was its original price?

CHAPTER 7

Simple and Compound Interest

Interest is the fee you pay when you borrow money, and it is a concept that dates back centuries. Surviving records show that interest was used in the ancient Babylonian, Greek, and Roman civilizations, and a moneylender is a key character in Shakespeare's *The Merchant of Venice*. Two types of interest plans are commonly used today—simple interest and compound interest. It is important to know how to calculate both.

What Is the Difference Between Simple and Compound Interest?

Simple interest is a one-time charge for the use of the amount of money that is borrowed. Compound interest is a plan in which interest is computed on a schedule, usually consisting of equal periods of time, and is based on the balance of the outstanding loan.

The modern financial world is built on credit. While borrowing money from a bank to buy a car, to buy a house, or to finance an education are well-known applications of using credit, consider the number of department stores and gas companies that also offer credit cards. Do you have a credit card from Sears, J.C. Penney, Macy's, Walmart, Mobil, or Sunoco? Credit cards are another way for companies to make money as they charge compound interest.

It is rare that one can get a loan with simple interest from a commercial establishment. It is not unusual, though, to get a loan with simple interest from a friend or relative, to help with a down payment for a car or a house, provided that the lender chooses to be so generous.

FACT

There are many references to the interest earned for lending money in literature aside from Shakespeare's *The Merchant of Venice*. Charles Dickens' *Bleak House*, *David Copperfield*, and *Hard Times*, and James Joyce's *Ulysses* are a few examples.

Computing Simple Interest

Simple interest is the product (the result of multiplying numbers together) of the amount borrowed (the principal), the rate of interest, and the amount of time for which the money is borrowed. As a formula, $I = P \times r \times t$, where I is the amount of interest, P is the principal, r is the rate of interest, and t is the amount of time. The rate of interest is stated with a time frame in mind (such as 6% per year), so it is important that the time period be expressed in units that correspond to the time frame of the rate of interest.

For example, let's say Charlie borrowed $2,500 from his parents to help him make the down payment on a new car. He agreed to repay the money to his parents in 2 years, along with 5% simple interest per year. Thus, in 2 years, he will be repaying the principal amount of $2,500 and paying simple interest at the rate of 5% per year. How much interest will Charlie owe his parents?

Answer: $I = \$2,500 \times 0.05$ (simple rate) $\times 2$ (time period) $= \$250$. Charlie will owe his parents $250 in interest, along with the $2,500 he borrowed, so he will owe them $2,750.

Try another example: What is the amount of simple interest charged for a loan of $500 for a 30-month period at a rate of 6%?

Answer: 30 months is 2.5 years ($^{30}/_{12}$). Therefore, $I = \$500 \times 0.06 \times 2.5 = \$75$.

Annual Percentage Rate versus Periodic Rate of Interest

Compound interest is charged periodically throughout the life of a loan. Familiar examples of compound interest include mortgage payments, car payments, and credit card payments, which are traditionally made monthly; the interest charges on these loans are also computed monthly. The interest earned in a savings account at a bank earns interest every 3 months (that is, quarterly). Payments on many government-sponsored student loans are also collected on a quarterly basis.

ESSENTIAL

The Annual Percentage Rate (APR) is the rate stated by a financial institution. The rate of interest for a given interest period is the APR divided by the number of interest periods in one year. The annual percentage yield (APY) is the effective annual rate after compound interest has been computed. The APY will be larger than the APR.

In all cases, the rate of interest is stated in terms of the *annual percentage rate* (APR). The rate that is actually used to compute the amount of interest for each period—the *periodic rate of interest*—is computed by dividing the APR by the number of interest periods in 1 year (this could be monthly, quarterly, etc.).

ALERT

It is important to know the meaning of the following terms that are used to describe the periods of interest in the language of borrowing and lending. *Annually* means interest is paid once per year; *semiannually* means interest is paid twice per year; *quarterly* indicates payment 4 times per year; *monthly* interest payments are made 12 times per year; and *biweekly* means interest payments are made 26 times per year (every 2 weeks).

For Example

Determine the periodic rate of interest for a car loan taken for 5 years at an annual interest rate of 5.8% compounded monthly.

Answer: Since there are 12 months in a year, the periodic rate of interest is 5.8% divided by 12, or 0.48333% per month. (The length of the loan, 5 years, has nothing to do with the periodic rate, because the periodic, or monthly, rate will be the same no matter how long a time is agreed to, by the lender and the borrower, for repayment of the loan.)

According to the website *www.creditcards.com*, the average annual rate of interest assessed on credit card balances as of March 9, 2013, was 14.95%. The monthly rate charged to an account is $\frac{1}{12}$ of the annual rate. In this case, that number is 1.24583333%. This rate will be multiplied by the balance due on the credit card on the day of the month specified in the contract between the user of the credit card and the company that issued it.

For example, if the balance on the credit card is $7,500 and you make a payment of $250, the new balance on your credit card will be $7,250 plus 1.224583333% of $7,250, or $7,340.32. The $250 payment you made only pays off $159.68 of the debt.

The Power of Compounding

The power of compounding is one of the greatest forces operating in the world of finance. With compounding, the value of money grows very quickly. A small number can grow into a very large number in only a few steps. The following problems should help illustrate this for you.

Examples of Compounding

Suppose you need to borrow $1,000. You have two choices: You can borrow the money from a friend who wants you to pay the loan back in 10 years with 5% simple interest, or you can borrow the money from a bank that will also charge 5% interest, but the interest is compounded annually. In each case, how much interest will you have paid to the lender when the loan has been completely repaid at the end of 10 years?

FACT

Albert Einstein is credited with saying, "Compound interest is the eighth wonder of the world. He who knows it, earns it . . . he who doesn't, pays it."

You owe your friend $1,000 \times 0.05 \times 10 = \$500$ interest, plus the $1,000 you borrowed, for a total of $1,500. That is the answer to the first question posed in this example.

To compute how much you owe the bank, you will calculate the amount of debt at the end of each year. (This will help explain the formula used to compute compound interest.)

At the end of the first year, you owe $1,000 for the principal, plus $1,000 \times 0.05 = \$50$ in interest, for a total of $1,050.

At the end of the second year, you owe $1,050 + \$1,050 \times 0.05 = \$1,102.50$. Note that applying the distributive property yields a pattern that makes this calculation very manageable:

$1,050 + \$1,050 \times 0.05 = \$1,050(1 + 0.05)$
$= \$1,050 \times 1.05$
$= (\$1,000 \times 1.05) \times 1.05$
(Recall $1,050 came from $1,000 times 1.05)
$= \$1,000 \times (1.05)^2$

See the table that follows:

End of Year	Amount Owed
1	$1,000(1.05) = $1,050$
2	$1,000(1.05)^2 = $1,102.50$
3	$1,000(1.05)^3 = $1,157.63$
4	$1,000(1.05)^4 = $1,215.51$
5	$1,000(1.05)^5 = $1,276.28$
6	$1,000(1.05)^6 = $1,340.10$
7	$1,000(1.05)^7 = $1,407.10$
8	$1,000(1.05)^8 = $1,477.46$
9	$1,000(1.05)^9 = $1,551.33$
10	$1,000(1.05)^{10} = $1,628.89$

Thus, at the end of 10 years, you will have paid the bank a total of $1,628.89. That is the answer to the second question posed by this example. To pursue this comparison a bit further, note that you are paying the bank $628.89 in interest (since the other $1,000 is just the return of principal). Subtracting the $500 in simple interest that you would have paid your friend reveals that you will pay $128.89 more by borrowing from the bank with the compound interest program.

This background explains the basis for the formula for computing the total amount, A, for principal and interest using compound interest. That formula is

$$A = P(1 + \text{periodic rate of interest})^{\text{number of interest periods}}$$

To reinforce the power of compound interest, suppose the loan you took from the bank in the previous example was for 10 years at 5% compounded semiannually. The periodic rate of interest would be 5% ÷ 2 = 2.5%, and the number of interest periods would be 20 (2 per year for 10 years). In this case, the amount of money due to the lender at the end of 10 years would be $1,000(1.025)^{20} = $1,638.62$, an additional $9.73.

Example: Suppose Jen borrows $5,000 from a bank at an interest rate of 6.24% compounded monthly for a period of 3 years.

The periodic rate of interest is 6.24% ÷ 12 or 0.52% per month and the number of interest periods is 36 (3 times 12). When the loan comes due, Jen will owe $5000(1.0052)^{36} = $6,026.42$. (Remember to change 0.52% to the decimal 0.0052.)

FACT

Accountants and finance people often refer to the "rule of 70." The rule of 70 gives an estimate of the amount of time it takes for money to double under compound interest. Divide 70 by the rate of interest to get the number of years before the money will double. For example, at 5%, the estimate is 14 years while at 6%, the estimate is just under 12 years.

Bank Accounts and Credit Cards

When you deposit money in a bank savings account, compound interest works for you, although the rates of return are very low. The money you deposit into such a bank account is, in reality, a loan you are making to the bank. The bank then lends that money to other customers and, in doing so, charges those customers a higher rate of interest than it pays you. That is how banks make their money. Some banks pay interest on checking accounts, but they require that a specified minimum amount of money be kept in the account. Savings accounts, which used to be the backbone of many families' savings programs, pay very little interest.

When examining the rates of interest on savings accounts, you should be careful to pay attention to the letters APR and APY. APR is the annual percentage rate before compounding occurs, and APY (the annual percentage yield) is the amount after compounding has already taken place for a year. For example, if a bank advertises that its APY is 0.84%, then the APR it is paying is 0.83% (the process for computing this is complicated). Consumers should know that the actual yield from an account earning compound interest is higher than the stated annual percentage rate. This is a consequence of the compound interest formula shown earlier in this chapter. There are some cynics who believe that when the financial institution cites the APY rather than the APR, the institution is trying to trick the consumer. A representative of the financial institution will argue that they are being transparent and are informing the consumer what the actual yield will be. Whatever the motivation, it is in the consumer's best interest to know the difference between these terms.

There are better savings plans than savings accounts (one example is certificates of deposit) that you should consider if you do not want to get involved in the stock market. By contrast, when you run an unpaid balance on your credit card, compound interest works against you.

ESSENTIAL

Be wary of all credit card offers that appear "too good to be true." Read all the fine print on the application for things like annual fees and the impact of missing a payment. The company issuing the credit card is in the business to make money. Some companies may include incredibly harsh consequences for late or non-payments.

Example: Alex has a credit card that charges an annual rate of 15.6% on balances not paid before the due date each month. The balance on his credit card for the past month was $4,000. How much interest was charged to his account this month?

Answer: The monthly rate for Alex's credit card is 15.6% ÷ 12 = 1.3%. 1.3% of $4,000 is 0.013 × $4,000 = $52.

Example: James has a credit card with an annual rate of 18%. His bill for the past month showed a balance of $6,500. He made a payment of $75. What is his new balance?

Answer: James paid $75 toward the balance of $6,500, leaving a debt of $6,425. The interest charged to the account will be 1.5% of $6,425 (1.5% is the monthly rate, which is found by dividing 18% by 12), so the interest charged this month will be 0.015 × $6,425 = $96.38. James will have a new balance, next month, of $6,425 + $96.38 = $6,521.38.

As you can see, the new balance on James's credit card will be higher next month than it was this month—even if he does not purchase anything with the card in the meantime! James has become a slave to debt, and unless he begins making larger payments right away, and does so for a long time, his enslavement will only grow deeper with time. This gives a clear warning of the financial danger you can get into by not paying attention to the amount of interest charged on your credit card(s) and the balances you are running. (The website *www.creditcards.com* cites a December 2009 study in which 36% of the respondents admitted that they did not know the rate of interest they were paying on their credit card balances.)

QUESTION

When do you work with simple and compound interest in your life?
Consider your bank accounts (savings as well as student loans) and car payments. What else did you add to your list of when you use math?

Exercises

Answer the following questions related to interest rates.

1. Alan borrows $1,500 from Rosemary with the intent to repay the loan in 3 years. If they agree on 8% simple interest, how much money will Alan need to repay?
2. Find the periodic rate of interest if the APR is 6% compounded quarterly.
3. Find the periodic rate of interest if the APR is 8% compounded monthly.
4. Quincy borrows $2,500 from the Adams Bank for 5 years with an agreement to repay the loan at 6% compounded quarterly. How much money will Quincy owe the bank?
5. Quincy chose not to borrow the $2,500 from the Hamilton Bank because it was charging 8% compounded monthly for the 5-year loan. How much did Quincy save by borrowing from the Adams Bank?
6. Arnie has a credit card that charges an annual rate of 15.6%. How much interest must Arnie pay this month on a monthly balance of $2,500?
7. Howard has a credit card that charges an annual rate of 10.8%. How much interest must Howard pay this month on a monthly balance of $8,200?

CHAPTER 8

Math in the Kitchen

There was a time in the not-too-distant past when converting pecks and bushels was part of the mathematics curriculum in America. Obviously, those days are long gone (unless your business happens to involve peppers or apples). In "deepening" the mathematics curriculum to take on more abstract concepts, it can be argued that some very important mathematics was lost—information such as the number of ounces in a pound and the number of teaspoons in a tablespoon. Why are facts like these important? Go to your kitchen and bake a cake. As it is often said by those who enjoy cooking and baking, "Cooking is an art, but baking is a chemistry lab."

Common Measurements

Starting with some of the smaller measurements, baking recipes use teaspoons, tablespoons, and cups as the basic units of measurement. These time-honored units, however, do not really give an indication of how one unit is related to another (this is unlike the metric system, where, for example, 1 liter = 1,000 milliliters). Some equivalencies worth knowing in the kitchen are

- 1 tablespoon (tbsp.) = 3 teaspoons (tsp.) = $\frac{1}{2}$ fluid ounce (fl. oz.) = $\frac{1}{16}$ cup
- 1 cup = 8 fl. oz.
- 1 pint (pt.) = 2 cups = 16 fl. oz.
- 1 quart (qt.) = 2 pt. = 4 cups = 32 fl. oz.
- 1 gallon = 4 qt. = 8 pt. = 32 cups = 128 fl. oz.
- 1 pound (lb.) = 16 ounces (oz.)

ESSENTIAL

Most people who cook will tell you that the most important application of mathematics in the kitchen is calculating the time needed to prepare each dish in a meal so that they are all ready at the same time to serve.

How do you convert among such units of measurement? Let's look at an example. If a recipe calls for $\frac{3}{8}$ cup of flour, how do you measure it? Because 1 tablespoon is equal to $\frac{1}{16}$ cup, you know that quantity will be easy to measure. Therefore, you decide to work in terms of sixteenths of a cup.

The fraction $\frac{3}{8} = \frac{6}{16}$. You know that $\frac{6}{16} = \frac{4}{16} + \frac{2}{16} = \frac{1}{4} + \frac{2}{16}$, so you can use a $\frac{1}{4}$ cup of flour and 2 tablespoons of flour to get the required amount.

ESSENTIAL

When measuring dry goods, be sure it is a level measure. You can run a straight edge, like the back of a knife, across the top of the spoon or cup to ensure the measure is level.

A 10-inch 1-crust pie calls for $\frac{1}{4}$ cup plus 3 tablespoons of lard.

How much lard is needed for a 10-inch 2-crust pie if the measurements are doubled?

Double $\frac{1}{4}$ cup to get $\frac{1}{2}$ cup and double 3 tablespoons to get 6 tablespoons of lard. Since one tablespoon is the equivalent of $\frac{1}{16}$ cup, the 6 tablespoons are the equivalent of $\frac{1}{4}$ cup (4 tablespoons) and 2 tablespoons. Therefore, $\frac{1}{2}$ cup plus 6 tablespoons is equal to $\frac{1}{2}+\frac{1}{4}$ cups, or $\frac{3}{4}$ cups, plus 2 tablespoons.

Dry Measurements versus Liquid Measurements

One major point of confusion that arises when working within the standard system is the difference between ounces and fluid ounces. The names look the same, and to some degree they are related.

The Difference

Ounces are a measure of weight, whereas fluid ounces are a measure of volume. The adage "A pint's a pound the world around" helps make sense of the use of the word *ounces* in both dry and liquid measures. Although that adage is true for water (and within reason for other liquids with a density similar to that of water), it does not hold true for a liquid such as crude oil, as is obvious every time a scene of an oil spill is shown in the news. The oil, being less dense than water, lies on top of the water rather than mixing with the water. For the same reason, vegetable oil in your kitchen will sit on top of the water in a pot unless you stir it in. On the other hand, a teaspoon of honey will sink in a glass of water because it is denser than water.

Because density is measured as a ratio (weight per unit of volume), multiplying by the volume gives the weight of the substance. What all of this means is that 16 fluid ounces (fl. oz.) of honey weighs more than 16 fl. oz. of water, and that 16 fl. oz. of water weighs more than 16 fl. oz. of vegetable oil.

FACT

A gallon of water weighs 8.34 pounds despite the saying, "A pint is a pound the world around."

Interestingly, there is no difference in volume between 1 cup measured with a dry-measure cup and 1 cup measured with a liquid-measure cup. What, then, is the reason for the different types of cups? The answer is convenience. Dry measurement is usually made by filling the container to the top and leveling it off with a knife (or some other straight-edged object). That is not easily done with a liquid. Cups for dry measurement usually come in sets with the volume they contain indicated: $\frac{1}{4}$ cup, $\frac{1}{3}$ cup, $\frac{1}{2}$ cup, and 1 cup. Need $\frac{3}{4}$ cup of flour? Use one $\frac{1}{4}$ cup and $\frac{1}{2}$ cup to get the required amount. By contrast, cups for liquid measure are transparent and graduated so that you can see the various measurements. You can measure $\frac{3}{4}$ cup of milk for your recipe directly—or even 5 ounces if that is what the recipe calls for.

ESSENTIAL

While a cup of dry measure has the same volume as a cup of liquid, measuring cups for liquids are calibrated in ounces and fractions of a cup for the sake of ease of use. There are a wide variety of recipes that call for different amounts of liquids while the call for dry measures is fairly standard.

Tablespoons and Cubic Centimeters

Kate and Russ live in Toronto, Ontario. Kate is an American citizen, born and raised in New York State, and she has a number of recipes from her childhood, as well as some she has collected as an adult. The measurements for these recipes are given in standard units, but the measuring cups she has accumulated in Toronto are in metric units. Kate has also shared, with her family and friends living in the United States, some of the recipes she has acquired while living in Canada. As you might expect, these recipes are written in metric units, and her U.S. family and friends cook with measuring cups in standard units. It is easy to convert rapidly between units in these two systems at any number of websites. Some of the more common conversions that you might need follow.

Volume	Weight
1 tsp. = 5 mL	1 oz. = 28 g
1 tbsp. = 15 mL	1 lb. = 450 g
1 cup = 240 mL	2.2 lb. = 1 kg
1 qt. = 950 mL = 0.95 L	
1 gal. = 3.8 L	

The recipe for Nestlé tollhouse cookies (from *www.verybestbaking.com*) calls for $2\frac{1}{4}$ cups flour, 1 tsp. salt, and $\frac{3}{4}$ cup brown sugar (among other ingredients). Convert these quantities to metric values.

- Flour: 1 cup = 240 mL, so $2\frac{1}{4}$ cups can be found by multiplying $2.25 \times 240 = 540$ mL
- Salt: 5 mL (one of the common conversions listed in the table)
- Brown sugar: $0.75 \times 240 = 180$ mL

Kate's mother-in-law, Rosemarie, gave her a recipe for Black Forest cake. The recipe calls for 75 mL water, 325 mL granulated sugar, and 4 mL baking soda (among other ingredients). Convert these quantities to standard units of measurement.

- Water: Because 1 tbsp. = 15 mL, divide 75 mL by 15 mL to get 5 tbsp. (or about $\frac{1}{3}$ cup).

- Sugar: 1 cup = 240 mL, so divide 325 by 240 to get 1.35 cups (that would translate into $1\frac{1}{3}$ cups).

- Baking soda: 1 tsp. = 5 mL, so use $\frac{3}{4}$ tsp.

QUESTION

A recipe given in standard units calls for 2.5 ounces of butter. How many grams of butter are the equivalent?
One (1) ounce corresponds to 28 grams so 2.5 ounces will correspond to $28 \times 2.5 = 70$ grams.

Making a Meal for More than the Recipe Calls For

Diane's Aunt Carol gave her a recipe for chocolate pie. The recipe makes one pie, but Diane needs to make three pies for the family reunion next week. Carol's recipe calls for $2\frac{1}{2}$ squares unsweetened chocolate, $3\frac{1}{2}$ cups milk, $\frac{3}{4}$ cup sugar, $\frac{2}{3}$ cup flour, $1\frac{1}{2}$ tsp. vanilla, 2 tbsp. butter, $\frac{3}{4}$ tsp. salt, and 1 egg.

How much of each ingredient will Diane need to make three pies? Clearly, she will have to multiply the amount needed for each of the ingredients by 3. The easiest of these computations is that she will need $1 \times 3 = 3$ eggs. As for the rest:

- Unsweetened chocolate: $2\frac{1}{2} \times 3 = 2.5 \times 3 = 7.5$, or $7\frac{1}{2}$, squares unsweetened chocolate

- Milk: $3\frac{1}{2} \times 3 = 3.5 \times 3 = 10.5$, or $10\frac{1}{2}$, cups milk. One cup of milk is 8 ounces, so $10\frac{1}{2}$ cups will be 84 ounces. One quart is 32 ounces, so Diane will need to purchase at least 3 quarts of milk for the recipe.

- Sugar: $\frac{3}{4} \times 3 = 0.75 \times 3 = 2.25$, or $2\frac{1}{4}$, cups sugar

- Flour: $\frac{2}{3} \times 3 = 2$ cups flour

- Vanilla: $1\frac{1}{2} \times 3 = 1.5 \times 3 = 4.5$ tsp. vanilla, or 1 tbsp. + 1 tsp. + $\frac{1}{2}$ tsp. vanilla

- Butter: $2 \times 3 = 6$ tbsp. butter. Because 1 tbsp = $\frac{1}{16}$ cup, 6 tbsp. = $\frac{6}{16}$ cup. Noting that $\frac{6}{16} = \frac{4}{16} + \frac{2}{16} = \frac{1}{4} + \frac{2}{16}$ shows that Diane will need $\frac{1}{4}$ cup butter + 2 tbsp. butter.

- Salt: $\frac{3}{4} \times 3 = 0.75 \times 3 = 2.25$, or $2\frac{1}{4}$, tsp. salt.

Knowing how to multiply the fractions will ensure that you have the correct amount of ingredients, and knowing the conversion factors will enable you to measure the materials more efficiently.

QUESTION

A commercial-size granola recipe calls for 22 quarts of rolled oats and 4 pounds of butter (among other ingredients). How much butter should you use if you will use 4 quarts of rolled oats?

Solve the proportion $\frac{4}{22} = \frac{amount\ of\ butter}{4}$, cross-multiply to get

$22 \times$ amount of butter $= 16$, and divide by 22 to get $\frac{8}{11}$ (which is approximately

0.73) lbs. of butter. In reality, use $\frac{3}{4}$ lb. butter.

Cooking Times

The U.S. Department of Agriculture website reports that approximately one-third of the population of the United States lives in high altitudes; a high altitude is defined as 3,000 feet (914 meters) or more above sea level). In high altitudes, the air pressure is lower, as is the moisture content in the air. This affects cooking times and temperatures. For instance, water boils at 208°F (97.8°C) rather than at 212°F (100°C). Consequently, most foods are cooked at a lower temperature and for a longer period of time. Should you be vacationing or moving to a place where the altitude is greater than 3,000 ft. (914 m), you will want to do some research on how altitude affects cooking. For instance, Martha Archuleta of New Mexico State University has a document online titled "High-Altitude Cooking" in which she says that at 5,000 feet, a 3-minute egg may take as long as 5 to 6 minutes. Hard-boiled eggs may need as much as 25 minutes to cook.

Older ovens may have much more variability in temperatures than newer ovens. That is, with an older oven, the temperature you set on the dial might not be the actual temperature inside the oven. Lower oven temperatures (say 300°F) on the thermostat might result in a lower temperature in the oven (like 275°F) while at higher thermostat temperatures (450°F) the actual temperature in the oven can be 475°F. An easy, and cheap, way to deal with this is to purchase an oven thermometer. You can use the oven thermometer to gauge how the temperature on the oven thermostat should be set to get the appropriate temperature inside the oven.

QUESTION

When do you use math in the kitchen?
Consider weighing foods and adjusting cooking times so that all the dishes are ready at the same time for the meal. What else did you add to your list of when you use math?

Exercises

Answer the following questions about math in the kitchen.

1. Kate is using Kristen's recipe for oatmeal peanut butter cookies. Kristen's recipe calls for $\frac{1}{4}$ tsp. of salt. Kate has to convert this to metric measurement. About how much salt does she need?

2. Brendon is using Kate's recipe for Black Forest cake. Kate's recipe calls for 175 mL of buttermilk. Brendon has to convert this to standard measurement. How much buttermilk does Brendon need?

3. A recipe for praline biscuits calls for $\frac{1}{2}$ cup of butter. The directions indicate that 2 tsp. of butter should be placed in a muffin tin. How many muffin tins can be filled with $\frac{1}{2}$ cup of butter?

4. A recipe for making calzones calls for $3\frac{1}{2}$ cups of flour to make four calzones. How much flower is needed to make two calzones?

5. A commercial recipe for pizza dough calls for 4 pounds $+ 5\frac{1}{3}$ ounces of water. How much water is this in terms of volume measurements?

CHAPTER 9

Budgets

CNNMoney.com reported that in 2012, the average debt carried by an American family that had at least one credit card was over $15,000. Use of credit to borrow for certain purposes is good (a mortgage, a college loan, or a car loan, for example), but borrowing can very quickly turn into burdensome debt, and too much debt is often hard to escape. Planning a budget is a good strategy to help you be sure that your expenses do not exceed your income. You will need to gather some information to help you plan a budget. This information includes an estimate of your annual income and statements about the expenses you can anticipate, such as the cost of housing (mortgage/rent and utilities), as well as an estimate of what you will spend on food, insurance premiums, and car payments, to name a few.

Projecting Gross Annual and Pay Period Income

You can begin your budgeting process by determining the amount of money you will earn in a year and estimating the amount of money you will earn each pay period. This will help you set a daily, weekly, or even monthly budget for yourself to keep you on track toward reaching your financial goals.

ESSENTIAL

Having a written budget is an excellent way to account for the money you earn and the money you spend. It is important to note that although the budget is not written in stone and completely inflexible, decisions to deviate from it should be made consciously and not on a whim.

Salaried

If you are a salaried employee, you know what your annual salary will be. You can divide your annual salary by 26 (assuming you get paid biweekly—that is, every other week) to determine the gross (before deductions) amount of each paycheck you will receive.

Example: Sarah is paid an annual salary of $54,700. Thus her biweekly income is $54,700 ÷ 26 = $2,103.85.

Hourly

If you are paid hourly for your work, you will need to estimate the number of hours you will work each week (including the number of hours for which you are paid at your base rate and the number of hours, if any, for which you are paid an overtime rate or at a rate different from the base rate). To calculate your weekly gross income, multiply your base rate by the number of hours you will work in a week (plus the number of hours you will be paid at a different rate times the number of hours you will work at that rate, if this applies to you). Your annual gross income is the product of this number and the number of weeks for which you are paid each year.

ALERT

Have a typical pay stub with you when you are trying to determine your monthly income. You should add bonuses and commission to the income line if these sources of income are regular and predictable.

Example: Alex earns $12.50 per hour for a 40-hour work week. He earns time and a half for working 6 hours beyond those 40 hours. His weekly gross income is therefore $40 \times \$12.50 + 6 \times 1.5 \times \$12.50 = \$500 + \$112.50 = \$612.50$.

Alex gets 2 weeks of paid vacation for his base pay (40 hours per week) each year. What is Alex's gross annual salary?

Alex is paid $612.50 per week for 50 weeks each year and $500 per week for the 2 weeks he is on vacation. His gross annual salary is $50 \times \$612.50 + 2 \times \$500 = \$30,625 + \$1,000 = \$31,625$.

Pay and Commission

Many salespeople earn a base salary plus a commission on the sales they make. Estimating salaries is a bit trickier in this case, because there is a need to estimate the amount of sales for each period (usually a month). As you might imagine, sales are seldom constant from month to month, and sales are dependent on the current economic climate.

Example: Janice earns a salary of $27,000 per year as a saleswoman. She also earns a 13% commission based on the gross profit that her sales bring in each month. In addition, Janice earns an extra $200 per sale if she sells more than 12 units per month. Data collected over a few years indicate that Janice can expect to sell 15 units each month (with an average gross profit of $2,500 per unit) for 8 months in the year. The other 4 months show average sales to be 9 per month (with an average profit of $2,100 per unit). Janice's estimate for her annual gross salary is

Salary:	$27,000
Commission for 8 months:	$8 \times 15 \times 0.13 \times \$2,500 = \$39,000$
Commission for 4 months:	$4 \times 10 \times 0.13 \times \$2,100 = \$10,920$
Bonus for 8 months:	$8 \times 15 \times \$200 = \$24,000$
Total = $100,920.	

She can estimate her semi-monthly (paid twice a month, or 24 times during the year) salary to be approximately $4,250.

Projecting Net Income per Pay Period

For a single person claiming no other deductions than for herself or himself, net (after taxes) income can be determined by taking 70% of the gross income per pay period. (The website *www.paycheckcity.com* can help you estimate the true amount of your net pay based on the tax rates for your state.) This is known as the 70% rule.

Say the three workers from the examples in the previous section—Sarah, Alex, and Janice—are all single and claim only themselves as a deduction on their tax forms. Let's estimate the net "take-home" pay per pay period for each of them. Look back and review the information you have about each wage earner's gross income, and then read on. After her employer withholds the legally required deductions for taxes, Sarah will take home approximately $2,103.85 \times 0.70 = \$1,472.69$ every other week. Alex will take home $\$612.50 \times 0.70 = \$428.75$ for the weeks when he works overtime and $\$500 \times 0.70 = \$350$ for the weeks when he does not. Janice will take home approximately $\$4,159.50 \times 0.70 = \$2,911.65$ twice each month. (If you want to perform these calculations for your own situation, remember that actual results depend on the state where you live, the number of dependents you claim, other deductions that you might be taking, and so on. Use the website to get a more accurate estimate for your particular situation.)

ALERT

Have as many of your typical monthly bills (housing, gas and electric, phone, insurance, etc.) with you when you are trying to determine your monthly expenses as possible.

Projecting Annual Expenses

When projecting your expenses for the year, you will need to determine which expenses are weekly, which are monthly, and which occur less frequently during the year. Gather this information and create a list with the headings "Weekly," "Monthly," and "Other." The list you create might look something like this:

Weekly		Monthly		Other	
Groceries		Rent		Auto insurance	
Gas		Car payment		Medical insurance premiums	
Coffee		Gas and electric		School and community taxes	
Lunch		Phone		Life insurance	
		Cable		Clothing	
		Child care			
		Entertainment			
		Savings			

You may have the records of what you spend on these items (in your checkbook, for example). Other items, such as what you spend for coffee or lunch each day, might very well be a mystery because you pay little attention to them. To calculate those expenses, it might help to get a large envelope (the kind that has a clasp and will hold an 8.5-inch × 11-inch sheet of paper without the paper being folded). For one "typical" month, get a receipt every time you buy something, write on the back of the receipt what you purchased, and place the receipt in the envelope at the end of the day. When the month is over, you will have a record of what you spent during the month.

ESSENTIAL

It is very easy to ignore the little purchases that are made day to day. Care must be taken to include these purchases (keep the receipts or write the amount of the purchases in a notebook as soon as is reasonably possible) because these little expenses can amass into a significant amount of money over a month.

Then you can go to your chart, write the amount you spend for each of these items on your list, and add the columns of weekly expenses and monthly expenses to get a pretty good estimate of what you spend at each of these intervals. You now have a sense of how your money is being spent each month. The items listed in the column headed "Other" may or may not be on a schedule. For instance, you probably do not purchase clothing on any type of a time schedule, but you do know that you will need to purchase clothing from time to time during the year. Estimate how much you think you will spend, and add that to the "Other" column. Insurance premiums and local taxes come due on predictable dates. You will need to save money for these expenses (as well as for vacations, special events, and emergencies that may occur). You might consider dividing the expenses from the "Other" column by 12 and making sure to put this amount into savings each month.

What's the Difference, and What to Do about It

If the amount of money you earn each month exceeds your monthly expenses (including budgeting for the "Other" expenses in your life), congratulations! You are on the right track and should be thinking about investing money for your future, especially for your retirement. If your expenses exceed your income, however, then you, like many other people, will have to make some tough decisions.

ESSENTIAL

When you spend more than you make, the totals in your budget will be negative, because when you subtract a larger number from a smaller number, the answer is negative. Let's say you brought home $3,114 last month, but you spent $3,788. To find the difference, you need to subtract, but the answer will be negative. $3,114 – $3,788 = -$674. A calculator will handle this easily, but if you're using pen and paper, you'll need to set up the problem a little differently. Put the larger number on top and the smaller number on the bottom. Then subtract, and remember, the answer will be negative.

Saving Money

If your monthly expenses exceed your income, you need to make some changes to get your finances in better "balance." One option to consider is earning more money. Is it possible for you to get a part-time job to help make up the difference between your monthly income and your monthly expenses? It is not uncommon for service industry stores to be looking for part-time help in the evenings and/or on weekends. You may be able to find such an opportunity that is compatible with your current work schedule and other commitments.

ESSENTIAL

Try to be as realistic as possible when making decisions on cutting expenses. While one big-ticket item might seemingly save a significant amount of money, if the elimination of the item will cause you to purchase a number of other items to make up for it, the reduction from the budget is not as significant. You may save more money by eliminating a number of smaller items that you do not need to replace.

A second option is to cut expenses. Although some expenses are beyond your control, there are some things you can do to reduce expenses. Grocery shopping with a list helps you resist making "impulse purchases"— that is, buying things you do not really need on a whim or just because they look good. Taking advantage of sales and comparative shopping can also save you money on groceries. Examine what you spend each week for lunch and coffee. Bringing your own lunch and/or coffee from home can save a lot of money. Carpooling can help you save money on gasoline. For example, if you carpool with one other person and drive on alternate weeks, you are paying for only half as much gasoline as you would have to buy if you drove to and from work alone each week.

You can save on gas and electricity in a number of ways. The prices for utilities are cheaper during non-peak hours, so running the dishwasher, washing machine, and clothes dryer during the off-hours can save money. Turning off the lights when you are leaving a room for a long period of time (as opposed to returning in 2 minutes) can save money, as can using more energy-efficient light bulbs. Adjusting the thermostat (somewhat cooler in the winter and somewhat warmer in the summer) is an effective way to reduce the electric bill. Your cable and entertainment bills can be lowered. If you must have cable television, choose the basic package and leave the premium options to others. Look in the paper or online for entertainment opportunities that have no cost or cost very little. Entertaining at home is cheaper than going out, so that is also an option to consider.

In the past decade, the cell phone has gone from a means of calling someone without having to look for a pay phone to a device that offers instant access to the world. However, data and texting plans, and the phones that use them, can cost more money than they are worth. You need to decide what is the bare minimum you can accept in a phone plan. Some people save money by no longer having a land-line telephone in their home and using their cell phone exclusively.

If you are having trouble paying your bills, the *least* desirable option is to get a credit card. The interest that mounts up on unpaid credit card debt is huge. If you must get a credit card, be aware of any annual fees, the rate of interest you will be charged on unpaid balances, and the extra expense you will incur each month.

Exercises

Answer the following questions about income and expenses.

1. Michael is single, claims just one deduction, earns $74,800 per year, and is paid biweekly. Use the 70% rule discussed in this chapter to determine the amount of his net pay each pay period.
2. Alysha is single, claims just one deduction, earns $18.50 per hour working a 45-hour week (with no overtime), and is paid biweekly. Use the

70% rule discussed in this chapter to determine the amount of her net pay each pay period.

3. Robin earns a salary of $32,000 per year and is also paid a commission of 15% of the gross profit for the sales she makes during the year. Last year, Robin's sales totaled $250,000 in gross profit. Robin is single, claims one dependent, and is paid twice a month. Use the 70% rule discussed in the chapter to determine the amount of her net pay each pay period.

4. Make your own list to determine where your money is spent.

CHAPTER 10

Taking a Trip

Gas prices in the United States have increased from below $2.00 per gallon in 2004 to above $4.00 per gallon currently. Even though the automobile industry has tried to create vehicles with more efficient engines to get better fuel mileage, taking a trip of over 200 miles can be expensive. Still, in a country that does not have a strong mass transit system outside of its major metropolitan areas, people are driving as much as ever before.

ALERT

How Far Away Is It and How Much Time Will It Take?

There are many popular websites for getting travel directions (among them are MapQuest, Google Maps, and Bing Maps). Of course, there are also GPS systems that you can purchase for your vehicle (Garmin and TomTom are the most popular brand names), and there are GPS apps that you can buy for your smartphone. Members of AAA can choose to get a TripTik to plan a route.

All of these products were designed to get you to where you want to go. Most of them provide you with options for your trip: Do you want to make the trip by traveling the least possible distance, by traveling the least possible time, or by avoiding toll roads? Do you prefer highway travel or a scenic route?

If you are traveling over familiar roads, you already know which route is shortest in distance and which route takes the least amount of time. When you are on a new adventure, it helps to plan how you are going to get where you are going and to have, at the very least, a map of the area where you are traveling in case you need to deviate from the route you originally planned. You never know when you start if there is road construction on your main route, an accident that will close the road for an extended period of time, or any other mishaps that might delay your travel.

While most people consider flying or driving as their main options, it might be the case that taking a bus or a train will fit your needs. You can ride the train in relative comfort—you can stand up and stretch your legs or possibly use the club car—and get a great view of the areas you pass through as you go. Bus rides also give you the opportunity to do some sightseeing while traveling that you might not get to do while you are driving. Whatever your destination and your means of travel, have a good trip.

How Much Will It Cost?

The primary cost of driving on your trip is gasoline. The average price of gasoline across the entire nation is often reported, but you should be more concerned with the price of gasoline in the area you will drive through and to. AAA has a website (*http://fuelgaugereport.aaa.com*) where you can get this information before you begin your trip. You can find statewide averages, as well as averages for selected markets. To compute an estimate for the cost of your trip, you will need to know the average number of miles your car gets per gallon of gasoline. (There is a difference between the mileage a car gets during highway travel and in local traffic, so it will help if you can get an average for highway driving.)

ESSENTIAL

Use your trip odometer to keep track of the number of miles you travel between stops to fill your gas tank. Each time you fill your tank, divide the number of miles on your trip odometer by the number of gallons needed to fill the tank. Be sure to fill the tank to get the best estimate for your gas mileage. (Do not, however, "top off" the tank, because this wasteful practice, which signs at many gas stations caution against, results in spills and contributes to ground and air pollution.)

Divide the number of miles you will drive—get this number from any one of the sources listed in the previous section—by the average number of miles per gallon your car travels. Multiply this number by the price per gallon of the gasoline you normally purchase.

Example: Brendon and Stacey are planning a trip to Michigan. The distance from their home to Michigan is 570 miles. Their car gets 24 miles per gallon for highway driving, and the average price of gasoline for their route is $3.587 per gallon. The approximate cost for gasoline for this trip, then, will be $570 \div 24 \times \$3.587 = \$85.19$.

FACT

The U.S. Department of Energy has a list of tips to help you maintain as high a mile-per-gallon average for your vehicle as you can at *www.fueleconomy.gov.*

Are We There Yet?

If you are driving long distances, you might set certain benchmarks to help pass the time. For example, you can calculate the distance between two major intersections so that as you leave one, you can estimate the time when you will arrive at the other. (This is particularly worth doing if traffic patterns such as rush-hour traffic will affect travel speeds—and thus travel time—along the route you are planning to take.) To compute driving time, divide the distance traveled (or to be traveled) by the speed at which you expect to travel.

For instance, say that you are driving on an interstate highway, and you pass a sign indicating that a rest area is 2 miles ahead and that the next rest area is 50 miles farther along the road. If you are driving at 70 mph, how long will it be before you reach the next rest area? Dividing 50 by 70 gives an answer of 0.714, but that is the number of hours. Multiply this number by 60, and you get an answer of 43 minutes.

Example: Your directions tell you that you need to drive 150 miles on the interstate, then to exit the interstate, and to drive another 75 miles on a local highway to reach your destination. The posted speed limit on the interstate is 65 mph, and the speed limit on the local highway is 40 mph. How long will it take to reach the destination? The time on the highway is $150 \div 65 = 2.31$ hours. The time on the local highway is $75 \div 40 = 1.875$ hours. The total travel time is 4.18 hours, or approximately 4 hours 10 minutes. Do not forget, however, that unless you exceed the speed limit during part of the trip, your average speed will be less than the legal limit. For this reason, you should allow for several minutes more than your calculation suggests.

FACT

You can estimate the time needed to drive to your destination by dividing the distance by the speed you will drive on that road.

Flying versus Driving

The airline industry has also been affected by the increased price of fuel. As a consequence, many airlines have made changes in how they do business.

Fares for flights have increased, and the number of flights has decreased. Many airlines charge for checking baggage and have introduced a scale for "preferred seating." There has been a reduction in the free amenities that were once common on flights. As a consequence of these changes, many travelers have stopped flying and are driving to their destinations instead. There are a number of factors to consider as you decide whether to fly or drive.

1. Cost—Compare the cost of a flight with the cost of driving.
2. Time—Trips in excess of 600 miles take a couple of hours in the air, whereas driving that far takes most of the day. How much time do you have to spend traveling?
3. Convenience—When flying, you go to the airport, submit to various security checks, board the plane, and then sit there for a few hours and read, write, listen to music, or possibly watch a movie. When you are driving, you are focused on the driving (although you may be listening to music or an audiobook while you do so). If you are a passenger in the vehicle, you have a few more options, but your ability to move about is still limited. Of course, when you drive, you can always pull off the road, get out of the car, and stretch.

ALERT

It is not an uncommon situation for airlines to overbook their flights and to ask for volunteers to take a different flight. If you have the flexibility to take another flight, it is reasonable for you to negotiate a travel voucher for a free flight at some point in the future. This could be an opportunity for free travel at some point in the future.

It becomes a matter of weighing these options to determine whether you should fly or drive.

Example: Diane and her husband are 800 miles from where their grandson lives. There is an airport 30 miles from where they live and an airport 25 miles from where their grandson lives. The cost of a round-trip ticket to see their grandson is approximately $400 per person, plus baggage fees and taxes. The total cost of the flight is approximately $900. Their alternative is to drive the 800 miles. Since they do not like to spend 12 hours in the car, they will break up the trip by staying in a hotel one night on the way to see their grandson and one night on the way home. The cost of the hotel is $120 per night. Meals for the road trip will total approximately $90 each way. Diane's car averages 30 miles per gallon on the highway, and the average price per gallon of gasoline on the route is $3.30. The cost of driving 1,600 miles is $(1,600 \div 30) \times \$3.30 = \$176$. The total cost for the road trip is $\$176 + \$240 + \$180 = \$596$.

Diane and her husband need to compare these transportation costs—$900 for flying versus $600 for driving. Of course, they should also take into consideration certain intangibles, such as the fact that if they fly, they will need to arrange transportation from the airport to their grandson's home, and, on the other hand, the mileage they will put on their car if they drive and the physical strain of being seated for that length of time. There is no universally correct answer here. Which mode of transportation you choose is a matter of personal preference. However, it is important to be aware of the expenses involved in making the trip with each mode of transportation.

QUESTION

How do you use math when you travel?
Consider the expenses of snacks and meals during your trip. If you are driving, you might also consider the expense of the mileage you put on the car (wear on the tires as well as the increased wear on the engine). What else did you add to your list of when you use math?

Exercises

Answer the following questions related to taking a trip.

1. Andrew and Robin are planning to take a trip to Charleston, South Carolina, a distance of 210 miles from home. They can get on the interstate highway in 5 minutes, and the average speed on the interstate is 65 mph. How much time will it take them to get to Charleston?

2. Andrew and Robin's car gets 28 miles to the gallon on the highway, and the average price of gasoline in their area is $3.40 per gallon. How much will they spend on gasoline during the round trip?

3. Kate and Russ are planning a trip to visit family members in Charlotte, a distance of 770 miles. To fly would cost them $680 in airfare. How much can they save by driving their car (which gets 38 miles to the gallon) if gas prices average $3.78 per gallon over the route they would take?

CHAPTER 11

Buying a Car

Buying a car can be a daunting task, but having a plan can help. Andrew Treanor of the Hendrick Automotive Group in Charlotte, North Carolina, says, "Know what you want to do with the car when the 'life' of the car comes to an end." Andrew's comments refer to the car owner's expectation of what he or she will eventually do with the car and at what point he or she will do it. Should you trade your car in after 2 or 3 years so it will still have a "new car" look, or should you keep the car until well after it is paid for? The answer to this question is a personal choice, but each driver should make that choice with a full understanding of the costs.

Buy or Lease?

Whether to buy or lease your vehicle is an important question—but you have to think about what you can afford. Inspired by the article "Driving Down Car Debt Commitment," the *Schenectady Gazette* ran an article with the headline "Buying a car should not bust a household budget." That was just the beginning of the *Gazette*'s advice. One of the themes running through this article—as indicated by the headline—is that people are spending more money on cars than they can really afford.

ESSENTIAL

One of the biggest questions a consumer must address with regard to getting a new car is whether it is more important to have a period of time without car payments or to always have a late-model car. While having a lease should guarantee that a minimum amount of money will be needed to be spent on maintenance, it also means that there will always be a car payment.

According to the article, when you are buying a car, you should pay cash whenever possible, but if you have to take out a loan, you should do so for a term of no more than 4 years. In addition, you should not pay more interest than is truly necessary.

The article also points out a few guidelines that you can use to stay on track with your car loan. One recommendation is that borrowers should keep monthly payments to no more than 20% of their monthly take-home pay and should apply the "20/4/10 rule." The 20/4/10 rule recommends making a 20% down payment on the price of the car, taking out a loan payable in no more than 4 years, and ensuring that the borrower's annual car payments and automobile insurance, combined, should not exceed 10% of her or his annual gross income. The authors also suggest that you consider buying a used car that is less than 3 years old, because most of the depreciation in the value of the car will have occurred, but the car itself will probably still be sound. Many automobile dealers offer "certified pre-owned cars" that include a warranty on the car.

Additional Advice

Mr. Treanor (you met him at the beginning of this chapter) also has some suggestions about buying or leasing a car. Car dealerships work on a quota system; that is, the dealership and the salespeople have quotas they need to meet each month and each year. The best times to buy a car are at the end of the month (especially the end of December, when annual *and* monthly quotas are due) and also at the transition point in mid-summer when all the new cars on the lot are last year's models and the dealership must make room for the new models that are being delivered.

He also suggests that you go to a site such as Kelley Blue Book (*www .kbb.com*) to do some research on the cost of the vehicle. You can enter the information on the vehicle and indicate any optional features that you would like for your new car and get the information on what the dealer is paying for the car and what the dealership is likely to charge for it in your area (you will need to give a zip code to start this process). You can then use this information to help you negotiate a better price for the vehicle that you want.

Mr. Treanor also has some advice on trade-ins and on leasing a car. You will get a better value for your car if you trade it in before the warranties on the car (such as the powertrain warranty) expire. He also suggests that you trade in the car when it still has a good deal of value. With regard to leasing, he notes that if you drive within the driving schedule of the lease contract, an investment for the first lease can be used on successive leases to keep you in a late-model car, similar to a down payment on the purchase of a new car. That is, when buying a new car, people will often trade in their old car. With leasing, you are trading in the old lease for the new lease. As the lease nears the end of its term, the dealership will be in a position to offer you a later-model car and keep the car payments close to the payments called for in the expiring lease. Speak with a dealership in your area to get specific details on the lease option.

ESSENTIAL

Auto dealerships earn most of their income from their service department, not the sale of the cars. You may find that dealers will offer competitive car loans with the hope that you will purchase a service contract with the dealership.

Shopping for a Car Loan

There are three major sources of loans for a car purchase: a local bank, a local car dealership (and the company that makes the car), and the Internet. The annual percentage rate (APR) for the loan—that is, the rate on which the compound interest formula will be based—can be found in the local newspaper. Some newspapers report these numbers each week, or you can read the ads from the different dealerships. You can also use the Internet to find rates. If you enter "automobile loans" into your Internet browser, you will get hits from any number of business concerns that offer such loans (usually from companies that pay the browser for a spot high on the list). Be wary! The deals offered by many of these companies include "fine print" that is very disadvantageous to the borrower if a there is ever an issue with a payment. It is wise to enter the name of a local bank into the browser as well, so that you can collect information from various lenders to help you make a better decision.

You should be looking for the rates for either new-car loans or used-car loans, depending on whether you are planning to purchase a new or a pre-owned car). You also want to note the number of months for which the money is being loaned; loans for new cars can generally be taken out for a longer period of time than loans for used cars. Bankrate.com (*www.bankrate.com*) reported that the average rate for a 60-month loan to purchase a new car on May 2, 2013, was 4.11% (the rate for a 36-month loan to buy a used car was 4.69%). Because this is an average, the rates for local lenders where you live might be lower. Take the time to find the best rate for you.

FACT

A loan with a lower number of years in its terms will result in lower amounts of interest being repaid but will have higher monthly payments. You need to keep your budget in mind when deciding on the number of years to take a car loan. You have to balance the other bills due each month versus the total amount of interest you will pay over the life of the loan.

Down Payments

Unless you are buying your first car (or need to purchase a second car while keeping your first car on the road), the down payment you will use for the purchase of your next car can often be the trade-in from your previous car. As in any purchase involving a money loan, the less you have to borrow, the better off you will be financially, because you will not have to pay back as much interest over the life of the loan.

As mentioned earlier in this chapter, it is advisable to have at least 20% of the cost of the new car (whether that be the trade-in or a combination of the trade-in plus cash) available for a down payment. Thus, if you want to purchase an $18,000 car, you should be able to put down $3,600 ($18,000 × 0.2).

Use the Kelley Blue Book website or the Edmunds website (*www.edmunds.com*) to get information on the value of your trade-in. Some of the information you will enter is somewhat subjective (such as your assessment of the condition of your car), but you will get a ballpark estimate of what you will be offered as a trade-in or what you can reasonably expect to sell the car for on your own, if you so choose.

The Car Payment

It is important to know what your budget will allow before you begin negotiations for a car loan. The Edmunds.com website (*www.edmunds.com/calculators*) has a calculator (found in the "New Car" tab at the top of the page if you enter Edmunds.com into your browser) that enables you to enter key information, such as what you would like your payment to be, the length of the loan, the tax rate where you live, and the amount of the down payment you plan to make. The resulting calculation shows the amount you can afford to spend on a new car. Choose the "What can I afford?" option, type in the dollar amount that works for your budget, click on the Go option, and answer the questions about your zip code, the interest rate, and the value of your trade-in and/or cash down payment. (The calculator will supply a rate for the zip code you indicate, but it may not be the same rate that you can negotiate with your lender.) You might want to play with this calculator to see how much you can afford to borrow based on the various interest rates that are available.

Please be aware that the numbers you will get when you do the calculations might not be exactly what is shown in this book. Market values are just that, they will change as supply and demand change. If the answers you get when you enter the data into the online calculators are within $500 or so of the results stated here, then you can be assured that you are entering the information correctly into the calculator.

For example, enter a payment amount of $500 with a 2.24% loan (taken from the Bank of America website on May 8, 2013), and $4,000 on a trade-in with $2,500 still owed on the trade-in. Consider three different options: a 5-year loan, a 4-year loan, and then a 3-year lease.

ESSENTIAL

There are many options for finding a loan for a car purchase. Whereas it used to be that you had to go to a bank to borrow money, there are a number of online services that offer competitive car loans. Many of the major automobile manufacturers have their own finance division for the purposes of making money for the company while offering the consumer competitive rates. Be on the lookout through newspaper, radio, and television ads for special interest rate incentives that car dealers offer.

Entering 60 for the term of the loan, in months, yields $26,200–$29,200 as a range of possible sticker prices, along with the suggestion that you not pay more than $27,700 fair market value, the value of a used car as determined by Edmunds or the Kelly Blue Book. If you enter 48 for the term of the loan, in months, the sticker price you can afford drops to $21,200–$24,200, with a suggested fair market value cap of $22,700. Reducing the term of the loan to 3 years (36 months) yields $16,000–$19,000 as a range of possible sticker prices, with a suggested fair market value cap of $17,500.

To Lease or to Buy?

Computing the amount of money you can spend on a lease is not as easy with this calculator. Use the information from the 60-month and 48-month loan options to compare what you can get for a lease option, and then go to the website AOL Autos (*http://autos.aol.com*). Use the suggested fair market value cap as the purchase price. Compare the lease to the 60-month loan, with a sales tax rate of 8% (adjust this for your area), a down payment of $1,500 (the difference between the value of your trade-in and what you still owe on it), $500 for other upfront costs (such as dealer fees), and an interest rate of 2.24%.

The lease term is 36 months, the monthly payment is $500, the security deposit is $100, you have to pay a cash deposit of $2,000, and there is no cash rebate. The savings tax rate is a minimal 1%, and state and federal tax rates (which you need to adjust for your situation) are 3.38% (for the purposes of this example).

Once you input all the information, the calculator will spit out your results. The average cost per year for the purchase is $4,019 and for the lease is $6,607. The monthly payments for the purchase are $462 and for the lease are $500. At a sales tax rate of 8%, $2,216 is due at the time of the purchase. Changing the fair market value cap to $22,700 and the term of the loan to 48 months, while leaving all other numbers the same, does not change the loan payments, but the sales tax due at the time of purchase is lower.

ESSENTIAL

The important point to remember is that you want to get a reliable car that fits your budget. Although you pay more overall when you extend the term of the car loan, the monthly payments will be lower. You want to pay as little in interest as you can, but you must do so within the constraints of your monthly budget.

Computing Monthly Payments

Whether or not you take the time to determine what price you can afford to pay for a car before shopping, you can still use the Edmunds calculator to determine your monthly payment.

Enter the price of the car you wish to buy in the Monthly Payment box on the website, and click on the Go button. Use $25,000 for the example.

Update the tax rate for your region if the stated rate is not correct (use 7% for the example), and update the Title and Registration value if the number stated is not correct (use $375 for the example). Include the amount in the box if you have a rebate or an incentive to include in the purchase ($0 for the example). Press the Next button. Enter the amount of your trade-in (for the example, $4,000 for the true market value and $2,500 for the amount still owed on the vehicle). Enter $0 for the down payment. Choose 60 months for the term of the loan and 2.24% for the interest rate. Click on the Calculate button. The monthly payment for the loan is $452, and after the term of the loan, $1,485 of the money paid will have been interest.

Work your way back to the beginning of the process for computing the monthly payment, and consider three different scenarios:

- You change the down payment from $0 to $500.
- You change the down payment from $0 to $500, and you change the term of the loan to 48 months.
- You leave the down payment at $0 and change the term of the loan to 48 months.

Your monthly payments (and the amount of interest you pay) for each of the scenarios are $443 ($1,457), $548 ($1,166), and $559 ($1,189), respectively. You need to choose the option that is best for you, given your budget and your needs.

Microsoft Excel

Excel has a couple of finance functions that can help with determining the amount of money you will spend on a car. The payment function (PMT) can calculate your payment given the APR (average percentage rate), the length of the loan, and the amount being borrowed (the present value of the money). For example, if you take a 5-year loan for $25,000 with an APR of 1.9%, enter =PMT(1.9%/12, 60, 25000) in the entry line. The result will be $437.10.

You entered 1.9%/12 because the interest is traditionally compounded monthly, and there are 60 months in 5 years. To calculate what you can afford to pay, use the present value (PV) function. If you think that you can afford $500 per month on a 5-year loan with APR 3.9%, enter = PV(3.9%/12,

60, 500). Here you'll find that you can afford a loan of $27,216.17 (this is the final value after down payment, taxes, and fees are applied to the negotiated price of the car).

Paying Off the Car Loan Early

Your monthly payment for your car loan is a combination of the amount of principal (the money you borrowed) and the amount of interest due on the principal. Should you decide to pay off the loan earlier than the time specified in the loan agreement, you do not do so by multiplying your monthly payment by the number of months still due on the loan. All you are required to pay (unless an early-payment penalty is included in your contract) is the amount of the principal that is still due.

ESSENTIAL

You do not need to wait until the end of the life of the loan to pay the loan off. If you have the good fortune to have extra income and want to pay off a loan, you may do so. If you choose to trade in your car while purchasing a new car, part of the agreement for the new car will include the dealership paying off the loan to the lender of the loan for the trade-in and will incorporate this cost into your new contract.

Here's an example to help illustrate: Jack and Diane took out a 5-year loan for $27,800 with an APR of 4.2%. They want to pay back the loan after 3 years. How can they do it?

Hofstra University has a calculator online for the Time Value of Money (*people.hofstra.edu/stefan_waner/realworld/tmvcalc.html*). To answer Jack and Diane's question, first compute the amount of each payment. The future value (FV) is 0, because the entire principal (all of the money borrowed) will be paid back. The present value (PV) is $27,800. You are trying to determine the payment, so leave this blank. The rate is 0.042 (change 4.2% to a decimal), m is 12 (12 payments per year), and $t = 5$ for the 5-year loan. Click on the Compute button on the PMT line to get $514.49. (The answer is negative because this is money that they will be paying out, not taking in.)

To determine the amount that Jack and Diane still owe, change the value of t to 2 (the number of years remaining on the term of the loan), and press the Compute button on the PV line. According to this calculator, Jack and Diane still owe $11,823.55. Compare this to the value $514.49 \times 24 = \$12,347.76$, the amount they would have paid if they had continued to make the car payments for 5 years.

Try another example: Chris and Kellie take out a 4-year loan for $23,900 at an APR of 5.1%. They find that they are in a position to pay the money back after a year and a half. How much do they owe on the loan? And how much money will they save by paying the loan off early?

First, determine that the amount of each payment is $551.48 by using the Hofstra website with $FV = 0$, $PV = 23900$, $r = 0.051$, $m = 12$, and $t = 4$. To get the early-payoff amount, change t to 2.5 (the number of years remaining on the term of the loan), and press the Compute button for PV. You will see $15,502.26, which is what they still owe on the loan. Compared to the amount they would pay if they let the loan run for its full term ($551.48 \times 30 = \$16,544.40$), Chris and Kellie will save $1,042.14.

QUESTION

How do you use math when you buy a car?
Consider the change in the cost of car insurance and the change in the amount of gasoline that you will use. What else did you add to your list of when you use math?

Exercises

Answer the following questions about getting a new car. (A reminder: the results are subject to market conditions. So long as the answers you get are near [within $500] of the answers in the answer key, you are doing the process correctly and are getting the latest market values.)

1. Kristen has decided to get a new car. In her area, interest rates for loans on the purchase of new cars range from 1.79% to 4.8%. Kristen will trade in her car, which has a fair market value of $3,200, and she owes no money on the car she will be trading in. Assuming that Kristen can

qualify for the lowest rates, what is the range of prices she can afford if she plans to budget $550 per month for a 48-month loan?

2. Laura and Tom decide to buy a new SUV with a fair market value of $38,650. The tax rate in their area is 7%. The value of their trade-in is $8,100, and they owe nothing on the car they will be trading in. Use the zip code 12866 on the Edmunds calculator to get the title and license fees. Determine the amount of their monthly payment under each of the conditions.

   a. 60-month loan, interest rate 2.4%, $750 down payment
   b. A 60-month loan, interest rate 2.4%, $0 down payment
   c. A 48-month loan, interest rate 2.4%, $750 down payment

3. Sumit has a 2011 BMW X3 with 32,000 miles on the odometer, a multi-disc MP3 player, and the premium package. It is in excellent condition. According to the Kelley Blue Book website, how much money should he expect to receive when he trades in this car? Based on the Kelley Blue Book values, how much more money could Sumit get if he sold the car himself? (Use the zip code for San Francisco, 94117.)

CHAPTER 12

Buying a House

Part of the "Great American Dream" is to own your own home. Being able to deduct mortgage interest when filing your taxes is also a wonderful part of owning that home. However, as the housing market collapse in 2008 showed, the practice of lending money to unqualified buyers left a number of people with houses they could not afford and drove the housing market (and the world economy) into one of the worst depressions since the late 1930s. Now, in 2013, the housing market is beginning to regain its footing, and interest rates for buyers of new homes are reasonable.

Shopping for a Loan

On May 9, 2013, the *USA Today* website (*www.usatoday.com*) reported that the mortgage lender Freddie Mac stated that 35-year mortgage rates rose from 3.35% to 3.42%, while 30-year loans dropped from 3.40% to 3.35%. It also reported that the rate for a 15-year mortgage rose 2.61% from a record low set 2 weeks earlier. Nationally, the average rate for a home mortgage was close to the November 2012 rate of 3.31%, and this was the lowest rate on record since 1971.

FACT

The word *mortgage* was probably coined in the fourteenth century. It's an Old French word that literally means "death pledge." But that doesn't mean that a mortgage will kill you. *Au contraire.* With a mortgage, the debt dies when it's paid. In other words, you can get out of a mortgage in one of two ways—by paying it off or by refinancing with another loan.

Just as in securing a loan for a new car, seekers of home mortgages have many sources to choose among, whether they are buying their first home or their fifth. Here again, it is important to check both with local lending institutions and with lenders who advertise on the Internet. (And always read all the fine print before you sign!)

ESSENTIAL

As with car loans, there are many financial institutions that offer house loans online with competitive rates. Taking the time to compare terms of various mortgages can save you a great deal of money over the life of the loan.

- Know Your Credit Score—Be aware of how those who lend money decide whether you are a good risk. A credit score in excess of 700 indicates that you should be a good candidate for a loan. Be sure to make your payments on any debt (such as credit card accounts) on time and never to miss a payment.

- Save Cash—There are a number of expenses that come with the purchase of a house. Fees for title searches, appraisals, home inspections, and credit checks, along with other closing costs, need to be paid without the aid of a loan. These expenses will run into the thousands of dollars. Lenders want to know that you have the ability to pay these expenses.

ALERT

When you buy a house with the assistance of a realtor, the seller's realtor usually—but not always—pays your realtor. Make sure you know whether this is the case for your transaction. If not, you will need to have cash to pay your realtor. Even though mortgage rates are more reasonable than they were before the "housing bubble" burst, there are stricter guidelines for lending money these days. Professionals who monitor the mortgage loan industry make the following recommendations to those who anticipate seeking a mortgage to buy a home.

- Stay at Your Job—Employment stability tells the lender that there is a reasonable expectation that your income will be not be interrupted and, indeed, a reasonable expectation that your income level will rise with pay raises and other work-related compensation.
- Know What You Can Afford and Get Preapproved for a Mortgage—The mortgage approval process can be lengthy (especially if you are a first-time buyer), so taking the time to get preapproved for a mortgage will give you an indication about how much the lenders believe you can afford to pay for a house. Keep in mind that owning a home involves significant expenses beyond mortgage payments. You will also have to pay community and school taxes, homeowner's insurance, heating costs and other utilities, as well as the usual expenses of day-to-day living, such as food, clothing, transportation, and the like. In addition, you are responsible for the upkeep of your home. Whether you hire someone to do the work or do it yourself, there is the matter of taking care of the yard, making any necessary repairs to your home (e.g., repairing or replacing a roof, plumbing and electrical

issues, painting), and replacing appliances as needed. If you are not aware of the many responsibilities involved in owning a home, speak to friends and family members who own a house about the general issues that arise. Since every house is different, no one can tell you everything about maintaining your home, but you can get an idea of the responsibilities you will take on as a homeowner—with no land-lord to call when a toilet springs a leak or high winds topple a tree onto the garage.

- Pay Down Debt and Avoid New Debt—Paying down your existing debt, while avoiding any significant new debt, is wise as you go through the process of getting approval for a mortgage. Remember also that a credit check will be run just before your closing on the house. Consequently, do not incur any new debt, such as buying new kitchen appliances, purchasing a new car, taking a great vacation, or anything of the kind, until *after* you have closed on the house and know that you can afford these expenditures. There are many stories of house sales falling apart just before the closing because a recent change in the buyer's credit report "scared" the lenders so much that they backed out of the deal.

FACT

The title search is used to determine whether there are any liens on the house or any restrictions on the use of the property. The party buying the house can contact a title company to do the search.

There are two types of loans on the market: fixed-rate mortgages and adjustable-rate mortgages (ARMs). A fixed-rate mortgage is exactly that; the rate you agree to when you sign the mortgage agreement is the rate you will pay for the entire life of the loan. An ARM offers a rate lower than the current fixed rate for a short period of time. Then the rate is adjusted on the basis of criteria established at the time the contract was signed. This rate becomes locked in for the remaining life of the loan. With an adjustable-rate mortgage, the borrower and lender are gambling on how the economy will change during the initial time frame. The borrower is hoping that mortgage interest rates will drop, and the lender is hoping those rates will rise. Also, people who intend to stay in the new home for a short period of time may take advantage of the adjustable-rate mortgage because they assume they will sell the home before the (possibly higher) fixed rate kicks in.

Down Payments and Points

Prior to the 2008 housing crash, it was possible to buy a home without making a down payment. As a consequence of the crash, there is now a debate among those in the housing industry about how much money should be required as a down payment. There was a time when a 20% down payment was expected. This level was later deemed unrealistic for lower-income buyers, and the rate was decreased. Some private lenders will accept 10% as a reasonable rate for a down payment. There are a number of government mortgage companies, such as the Federal National Mortgage Association (Fannie Mae) and the Federal Housing Association, that offer mortgages with down payments less than 10%.

ESSENTIAL

There are a number of government programs that are meant for first-time homeowners. The requirements to qualify for these loans are very specific. Contact the Department of Housing and Urban Development to see if you qualify.

Lending institutions require that borrowers who borrow with a low down payment take out private mortgage insurance (PMI) in case they default on the loan. Private mortgage insurance must be paid for at least 12 months and until the home reaches at least an 80% loan-to-value ratio (the ratio of the balance on the loan to the value of the home). For example, if the mortgage is for $150,000, the loan-to-value ratio would be 80% if the appraised value of the home were $187,000. A loan-to-value ratio represents division of the amount of the loan by the appraised value of the home, so the appraised value of the home is found by dividing the amount of the mortgage by 80% (in the decimal form 0.8). Because $150,000 ÷ 0.8 = $187,000, the value the home must be at least $187,000 in order for the borrower not to have to carry private mortgage insurance (PMI). Thus borrowers who need to make a down payment of less than 20% (because $1.00 - 0.80 = 0.20 = 20\%$) have to carry PMI until their monthly mortgage payments gradually reduce the balance remaining on the mortgage to less than 80% of the value of the home.

QUESTION

How much money must a buyer have as a down payment for a home that costs $210,000 if the buyer does not want to carry private mortgage insurance?

The minimum the buyer must have as a down payment is 20% of $210,000, or $42,000.

Many institutions charge the borrower a percent of the money being borrowed at the time of the closing. This fee, which is usually referred to as points, is negotiable. For example, on a mortgage of $150,000, a loan with 2 points due at closing means that the borrower must pay 2% of the $150,000 loan (or $3,000) at the time of the closing. A lender may offer a lower rate on the mortgage for a higher point value at the time of closing. Ask about this, and run the numbers to see which option is best for you.

FACT

Escrow: Some lending institutions require that the borrower make monthly deposits into a separate account called an escrow account to ensure that money is always available to pay property taxes and home insurance premiums. This account usually pays the borrower a nominal rate of interest. Whether such an escrow account is required may be negotiated when the terms of the mortgage are drawn up.

The Mortgage Payment

There are a number of calculators that can be accessed online to help you compute your mortgage payment. The website *www.mlcalc.com* has a calculator that enables you to enter all the information—sale price, percent down payment, term of the mortgage, property tax, annual property insurance cost, and PMI (in case the down payment is less than 20% of the purchase price)—that you need to determine the mortgage payment as well as the escrow payment. The website *www.MortgageCalculator.net* enables you to enter the amount of the loan and the term of the loan (including the option of choosing an adjustable-rate mortgage) and then computes the monthly payment, the total amount repaid, and the total interest paid on the loan.

ESSENTIAL

With interest rates in the middle of 2013 available at 3% for some buyers and as much as 6% for others, there is a wide range of money that will be paid back over the life of the mortgage. A 3% mortgage for 15 years will result in the homeowner repaying approximately 1.4 times the amount borrowed while a 30-year, 6% mortgage will have the homeowner repaying more than twice the amount borrowed. Is this a bad thing? Not really, when you consider that the homeowner has had access to that money (in the form of the home) for an extended period of time.

The examples for the rest of this chapter will be done with the calculator from Hofstra University (*people.hofstra.edu/stefan_waner/ realworld/tmvcalc.html*). The information you will need to enter includes FV, PV, PMT, r, m, and t. In the context of mortgage loans, the meanings of these abbreviations are as follows:

- FV—future value—the amount that will be in the account at the end of the loan. At the end of the loan, you should owe no money, so this number is 0.
- PV—present value—the amount that is in the account today (or the amount that you still owe today). On the day that you sign the papers

for the loan, this number will be the full amount you are borrowing (which is usually not the full purchase price of the house). To find the present value, subtract the down payment from the purchase price).

- PMT—payment—the amount of the monthly payment for principal and interest (P&I). The monthly payment is what you are trying to determine, so leave this box blank.
- r—the annual rate of interest expressed as a decimal.
- m—the number of times per year that payments will be made on the loan.
- t—time—the number of years for which the loan is being taken out.

Mortgage Examples

Chris and Diane buy a home for $250,000. They pay 15% as a down payment and finance the rest with a 15-year, 3.45% mortgage. Determine the amount of their mortgage payment.

First, Chris and Diane make a down payment of $37,500 (0.15 × $250,000), which leaves $212,500 for them to finance. Using the Hofstra online calculator, leave the PMT box blank and enter FV = 0, PV = 212500, r = 0.0345, m = 12, and t = 15. Click on the Compute button next to the box for PMT to learn that their monthly payment will be $1,513.91.

Example: Mike and Aileen buy a home for $275,000. They qualify for a 30-year mortgage at 3.75%. They are debating whether to make a 10% down payment or a 15% down payment on the purchase. Compute the monthly payment for each case, and compare their options.

A 10% down payment ($27,500) leaves a balance of $247,500. With PV = 247500, r = 0.0375, m = 12, and t = 30, the monthly payment is $1,146.21.

A 15% down payment ($41,250) leaves a balance of $233,750. With PV = 233750, r = 0.0375, m = 12, and t = 30, the monthly payment is $1,082.53.

Spending an extra $13,750 today saves Mike and Aileen $63.68 a month. There are two ways to think about this. On the one hand, they can use the $13,750 to fix up the house without having to borrow more money now, or they can invest it, which makes sense if they can get a higher return than 3.75%. On the other hand, they will pay an extra $63.68 for 360 months—a total of $22,924.80. Consider what you would do if you had to make this choice.

ALERT

When you close on the purchase of a house, bring your checkbook. Unless you've negotiated something different with the seller, you'll have to lay down some cash before the deal is done. Here are some of the costs you might be expected to cover at closing. *Processing fees* cover the administrative costs of processing your loan, including expenses incurred in checking your credit report, the lender's attorney's fees, document preparation costs, and so on. *Points* are a one-time charge you may pay to lower your interest rate over the life of the mortgage. An *appraisal fee* is charged when your lender needs to establish the actual value of the home. The *title fee* covers the cost of ensuring the home belongs to the seller, a process called a title search. This fee may also include title insurance, which protects the lender against an error in the title search.

Example: Meghan and Will decide that they would like to buy a condominium. They examine their budget and determine that they can afford a mortgage payment of at most $1,100 each month and that they can afford to make a $10,000 down payment on the house they buy. If the best interest rate for a 25-year mortgage in their area is 3.9%, how much money can they afford to spend on a condo?

The payment is known to be $1,100. It is the present value that Meghan and Will do not know. Set PMT = 1100, $r = 0.039$, and $t = 25$. Click on the Compute button on the PV line to determine that the maximum mortgage they can afford is $210,594. Add the $10,000 down payment, and the total cost of the house can be $220,594. Note that the $10,000 down payment is less than 5% of the purchase price. If Meghan and Will cannot get a mortgage that allows such a low rate of down payment, they will need to consider buying a house that is worth less than $220,594. Otherwise, they may have to postpone their purchase of a home until they can make a larger down payment.

Paying Off the Mortgage Early and Home Equity Loans

Once you own a home, there are a few things you'll need to consider. When the time comes for you to sell the home, how much will you owe on it and how much of the sale price will you get to keep? If you plan to stay in the home, there is the question of how much of the original mortgage you have repaid. Home equity mortgages represent an opportunity to borrow (from any bank that makes such loans), and put to any use you choose, some of the money you have repaid on your home mortgage. Home equity mortgages may also offer tax advantages to you.

Examples

In the previous section, Chris and Diane took a 15-year loan on their house and have a monthly payment of $1,513.91. Twelve years later, they sell their house for $280,000. Their realtor gets 7% of the sale price. How much money do Chris and Diane get after the mortgage is repaid and the realtor is paid?

ESSENTIAL

The Real Estate Settlement Procedures Act (RESPA) requires that all lenders give the borrower a Good Faith Estimate (GFE) of all the fees associated with the loan. This way, you'll have an idea of how much you'll pay in closing costs. However, a Good Faith Estimate is just that—an estimate. The actual fee may be slightly lower or higher.

Chris and Diane still have 3 years of payments to make on the mortgage. (This is assuming that the sale of the home occurs exactly 12 years from the day of the first mortgage payment. In reality, Chris and Diane might have 34 payments still to make, in which case the time would be $\frac{34}{12}$, or 2.867.)

Enter 1510.79 for the payment, 0.0345 for r, and 3 for t. Calculate the present value to be $51,598.11, the amount they still owe on the original mortgage of $212,500. The realtor's fee is 7% of $280,000, or $19,600. Chris and Diane keep $280,000 − $51,598.11 − $19,600 = $208,801.89.

Example: Mike and Aileen (also from the last section) decided to make a 10% down payment. Ten years later, their house is assessed at $328,000. Their bank will give them a home equity loan for 80% of their equity. (Their equity is the difference between what their house is worth and the amount they still owe on their mortgage.) Mike and Aileen would like to expand their garage to hold two cars and add a guest room above the new garage. How much money can they get by taking out a home equity loan for this purpose?

Compute the amount they still owe by entering PMT = 1146.21, r = 0.0375, and t = 20 (for the 20 years still left on the mortgage). Compute the present value to be $193,326.46. The difference between the value of the house and the amount that they owe is $328,000 − $193,326.46 = $134,673.54. The amount of the home equity loan they can get is 80% of this number, or $107,738.83.

QUESTION

How do you use math when you buy a house?
Consider the change in the cost of heating and cooling the house, the change in insurance, and the change in the cost of getting to work. What else did you add to your list of when you use math?

Exercises

Elizabeth decides to purchase her first house for $215,000. She will make a $15,000 down payment and finance the rest of the purchase by taking out a 4.1%, 30-year mortgage.

1. How much money will Elizabeth need to borrow?
2. What is the amount of her monthly mortgage payment?

Five years after buying this house, Elizabeth sells it for $250,000.

3. How much does she still owe on the original mortgage when she sells?

4. The realtor gets 6% of the selling price. How much is the realtor paid?
5. How much money does Elizabeth receive from the sale of the house?

Elizabeth uses the money from the sale of her first house as a down payment on the purchase of a new house for $310,000. She takes out a 25-year mortgage with an APR of 4.4%.

6. What is the amount of money she needs to borrow?
7. What is the amount of her monthly payment on the new house?

CHAPTER 13

Saving for Retirement

"Start saving as early as you can," says Stacey Treanor, CFP®
(Certified Financial Planner). "At the beginning of your career,
retirement may be the last thing on your mind, but it's important to begin saving even if it is just a small amount. By saving
at an early age, you are able to take advantage of the power
of compounding interest. If you wait until the end of your career to begin saving, you have to make up for lost time by
saving at a higher rate. In addition, there are tax benefits associated with saving for retirement, and you'll want to check
with your employer to see if they make matching contributions to a retirement plan. For example, a common employer
matching formula is 50% up to 6% of eligible compensation.
In this scenario, if an employee contributes 6% of his or her
pay to the retirement plan, the employer is also contributing
3%. If your employer matches contributions, try to contribute
at least enough to receive the full employer match. If not, you
are essentially turning away free money."

IRAs and 401(k) Plans

Two savings plans that allow a person to save money for retirement without paying taxes on that money until retirement are individual retirement accounts (IRA) and 401(k) plans. Any individual can create an individual retirement account (IRA) through a bank or another financial institution such as a brokerage firm or a mutual fund company. The vehicle for retirement saving known as the 401(k) is available only to people who work for a company that offers such a plan. Treanor's advice about the benefits of saving for retirement early applies to everyone, whereas her comments about the contribution of matching funds by an employer are for those employees to whom a 401(k) is available.

If a 401(k) plan is available to you, here are some things you should consider:

ESSENTIAL

While the issue of whether Social Security can continue to exist has been heatedly debated in Washington, D.C. and elsewhere, this fund was never meant to be the sole means of supporting retirees. Add to this fact that since the inception of Social Security during the Franklin Delano Roosevelt administration, the life expectancy for Americans has risen more than 15 years. This has put a great strain on the funds in the Social Security system.

- Money contributed to your account by your employer may not stay with you if you leave the company. Investigate the amount of time you must be employed by the company before you become vested—that is, how long it will be before you can keep the money if you leave the company.
- As Treanor stated, it is fairly common for an employer to match 50% of what you contribute (up to 6% of your annual salary) to your 401(k) plan. What are the specifics for your company?
- Money put into a 401(k) is not taxed until you begin to withdraw it, so it grows more rapidly than an investment where gains are taxed each year. Another advantage is that there is a reasonable expectation that your tax situation will benefit from your making contribu-

148

tions to a 401(k), because you will not be withdrawing money from it at the same rate at which you were being paid while you worked. It is also possible that you will be in a lower marginal tax bracket in retirement than while you were working.

- As of 2013, the maximum you can contribute to a 401(k) in a year is $17,500 if you are 49 years old or younger. This limit rises to $23,000 if you are older than 49.
- You may *not* access the money in the account (except in the form of a loan that must be repaid in its entirety) until you reach 59.5 years of age, and you *must* begin to withdraw your money by age 70.5.
- You have limited ability to determine the way the money you have contributed is invested. Many times, the company will invest the money in itself. This is rather tricky when you consider how many companies go bankrupt.

FACT

The 403(b) is a tax-advantaged retirement savings program available to public education employees and some nonprofit employers that is like the 401(k) for the private sector. IRS Publication 4484 has information for public sector employees.

Like the money you contribute to a 401(k), the money invested in an IRA is deducted from your taxable income, and you do not begin paying taxes on this money until you begin to make withdrawals from the account. You may not borrow money from a traditional IRA. You may, however, roll the money over from a traditional IRA into what is called a Roth IRA. The Roth IRA has provisions for borrowing money for things like buying a house, paying for education, and medical expenses. The maximum annual contribution to an IRA is $5,500 if you are 49 or younger and $6,500 if you are older than 49.

ESSENTIAL

Simple Annuities

In setting up an IRA, you have the option to invest in the financial markets through a brokerage house or by investing directly in a mutual fund or funds. When you invest, the rate of return on the money invested can be estimated on the basis of the past performance of the products selected. Whether to invest in the stock market, the bond market, or a combination of the two is something that you should discuss with a certified professional. Such discussions are beyond the scope of this book. Other options for investment include money market funds, certificates of deposit, government bonds, and fixed-rate annuities.

A *simple annuity* is one in which the deposit into the account occurs in the same time frame in which interest is credited to the account. For example, if the account pays interest compounded monthly, deposits into the account are made each month. In terms of compound interest, the simple annuity is akin to having a separate savings account for each of the deposits made into the account. For example, suppose that Reza, age 25, decides to deposit $600 each year into a simple annuity that pays 1.51% compounded annually, with the first deposit being made on her 25th birthday. Furthermore, assume that Reza has no plans to access this money until she is 65. How much money will this annuity grow to? Look at the accompanying table.

$600 Deposited on Birthday #	Number of Years to Age 65	Amount to Which Deposit Has Grown at Age 65
25	40	$600(1.0151)^{40} = \$1092.71$
26	39	$600(1.0151)^{39} = \$1076.45$
30	35	$600(1.0151)^{35} = \$1013.82$
40	25	$600(1.0151)^{25} = \$872.71$

This table compares the values to which a single deposit of $600, made at four different ages (25, 26, 30, and 40), will grow by the time the depositor reaches 65 years of age. To see what happens when Reza deposits $600 on every birthday over the 40 years, use the calculator from Hofstra (*people .hofstra.edu/stefan_waner/realworld/tmvcalc.html*).

Prior to making her first deposit, Reza had no money in the account, so the present value (PV) of the account is 0. Each payment (PMT) is $600. The rate of interest is 1.51%, so enter 0.0151 for r. Interest is compounded once per year, so $m = 1$. Forty years pass from the creation of the account until Reza is ready to withdraw the money, so $t = 40$. Click on the Compute button beside the future value (FV) box to get $32,691.72 (ignore the negative sign). Is that enough money to retire with? No, absolutely not. However, it does suggest the impact of the annuity. Reza made 40 payments of $600 each for a total of $24,000 invested, and she earned over $8,000 in interest.

A more reasonable example might be that Reza makes monthly deposits of $50 into an IRA that pays 1.51% compounded monthly. (This is more reasonable in that it is usually easier for a person who is just entering the work force to give up $50 at one time, (especially if she takes $25 from each of two paychecks in a month for this purpose), than to come up with $600 all at once.) Reza continues this pattern of contributing to her IRA for 5 years. On her 30th birthday, she decides that she can double the size of the monthly payment. At age 40, she increases her monthly payments to $450. Finally, at age 50, she increases the amount of the payment to $540 per month. (This keeps her within the current guidelines for maximum deductions without incurring a tax penalty.) How much is in the account when Reza reaches age 65?

To do this problem, it is necessary to investigate the amount of money that is in the account at each of the points when Reza changes the amount of her contribution.

1. Stage 1 (ages 25 to 30): Reza starts with no money in the account (PV = 0) and pays $50 per month (PMT = 50) into an account that pays 1.51% ($r = 0.0151$) compounded monthly ($m = 12$) for a 5-year period ($t = 5$). Compute future value (FV) to be $3,114.12.
2. Stage 2 (ages 30 to 40): Reza has $3,114.12 in the account (PV = 3114.12), and the payment changes to $100 (PMT = 100). Ten years later ($t = 10$), she has (FV) $16,565.97 in the account.

3. Stage 3 (ages 40 to 50): The present value is now $16,565.97, and the payment becomes $450. Reza will have $77,515.06 in the account.

4. Stage 4 (ages 50 to 65): The present value is $77,515.06, the payment is now $540, and the time frame is 15 years ($t = 15$). Reza will have $206,217.39 at age 65.

Is this enough money for retirement? Probably not, when you consider inflation and the lifestyle Reza may want to enjoy. But combined with Social Security and her other savings and investments, it can help form a basis for a secure retirement.

How Much Will I Need?

It's time for you to break out the crystal ball and predict the future. How much money will you need at retirement? Here are some very important questions to consider:

ALERT

According to an article in the March 25, 2013 issue of *MoneyWatch*, the "average career worker" will need 11 times his or her average salary (in savings, assets, and benefits) to be able to retire.

- Housing Expenses—If you own a home, can you pay the mortgage off before you retire? Can you have things like a new (or relatively new) roof, furnace, and water heater installed so that you won't have any unanticipated major expense for the house after you are retired?
- Health Insurance—What do you anticipate having to pay for health insurance and medical expenses? In 2013, the Health Care Plan for America is being ironed out. Since it is such a highly political issue, it is difficult to predict what the cost of health care in America will look like in 10 years.
- Lifestyle—Do you plan to travel? If so, by what means? Will you drive, fly, or take a cruise? How often will you travel? Will you make one major trip each year or a couple of smaller trips during the year? Con-

sider the costs of transportation, lodging, and meals. Do you like to go out to eat? Do you like to go to the movies or attend plays and musical performances? What are the things that you want to be sure you can do in retirement while your health and your money allow for it? Determine an amount that will represent your expected entertainment expense.

- The Basics—When you are not traveling, you will need to consider what your food and beverage budget, your utility budget, and the amount you pay for insurance (health, auto, renter's, etc.) will be.
- For How Long?—How many years do you anticipate that you will be living on the money you have saved for retirement? (At what age do you expect to retire? How is your health? Does longevity "run in your family"?) These are difficult questions to ask (and maybe even more difficult to answer), but thinking carefully about them is an important part of planning for retirement.

I'm Retired—How Much Can I Withdraw?

You have made it to retirement (congratulations!) and are ready to begin the next phase of your life. You are getting ready to set up a budget to govern your retirement lifestyle. At this stage, you are probably not making any contributions to a retirement account (although there may be cases where that does occur). Now you have to determine how much cash is available to you in your IRA, 401(k), savings account, and other investments. And you need to determine how much money you can withdraw from your retirement account each month.

Examples

Marcel, age 65, has a retirement portfolio of $350,000 invested in an account that pays 1.6% interest compounded monthly. If Marcel wants this money to last for 20 years, what is the maximum amount of the monthly withdrawal he can take from the account?

Using the Hofstra calculator, the present value of the money in the account is $350,000 (PV = 350000), and the value of the account after 20 years ($t = 20$) is 0 (FV = 0). With $m = 12$ and $r = 0.016$, press the Compute button next to the PMT box to determine that Marcel can withdraw $1,705.05 from the account each month.

Example: William, age 62, has a retirement portfolio of $650,000 invested in an account that pays 1.75% interest compounded monthly. He plans to withdraw $4,500 a month from the account for 10 years. After that, he will "settle down" and use the balance of his account by making equal monthly withdrawals for 15 years. What amount will he be able to withdraw each month during the last 15-year period?

The present value is $650,000 (PV = 650000), $r = 0.0175$, $m = 12$, and $t = 10$. The amount of each withdrawal for the 10-year period is negative $4,500 (–4500), because the money is causing the balance in the account to drop during this time. The balance in the account after 10 years (the future value of the account) is found by pressing the Compute button next to the FV box. The amount left is $184,548.72.

To determine the amount of the monthly payments after that, set PV = 184548.72, FV = 0, and $t = 15$. Press the Compute button next to the PMT box to find that he will be able to withdraw $1,166.46 a month during the last 15 years.

QUESTION

How do you use math when you save for retirement?
Consider how you plan to travel and what new expenses you may incur as a result of having more free time. What else did you add to your list of when you use math?

Exercises

Jack and Diane, each age 28, open IRA annuities that pay 1.4% compounded monthly and decide that they will put $120 per month into their respective annuities beginning with their 28th birthdays.

1. How much money will be in the account when Jack and Diane are 35? Beginning with their 35th birthdays, they increase the amount of their monthly contributions to $400.
2. How much money will they have in their accounts at age 50?

Beginning with their 50th birthdays, they make monthly deposits of $540 into their accounts.

3. How much does each have in her or his account at age 65?

Jack and Diane roll over their IRAs into one account that pays 4.2% compounded monthly. They plan to make equal monthly withdrawals from the account for the next 25 years.

4. What is the amount of each withdrawal?

CHAPTER 14

Filing Tax Returns

Albert Einstein once said, "The hardest thing in the world to understand is income tax." If you agree with Albert Einstein, you aren't alone. Millions of people struggle to complete their income tax returns. Every state has its own income tax laws and forms, so this chapter focuses only on the federal form 1040EZ.

Which Form Is Appropriate?

Form 1040EZ is the easiest of the federal tax forms to complete. However, depending on your personal situation, it may not result in the best tax outcome for you—that is, you may be entitled to deductions on the 1040 that exceed the standard deduction allotted to you on the 1040EZ. According to the Internal Revenue Service document "Topic 352: Which Form—1040, 1040A, or 1040EZ?" (*www.irs.gov/taxtopics/tc352.html*), you could use Form 1040EZ to file your tax return for the year 2012 if you meet all of the following criteria:

- Your filing status is single or married filing jointly.
- You claim no dependents.
- You, and your spouse if you are filing a joint return, were under age 65 on January 1, 2013, and were not blind at the end of 2012.
- Your income is only from wages, salaries, tips, taxable scholarship and fellowship grants, unemployment compensation, or Alaska Permanent Fund dividends, and your taxable interest was not over $1,500.
- Your taxable income is less than $100,000.
- Your earned tips, if any, are included in boxes 5 and 7 of your Form W-2.
- You do not owe any household employment taxes on wages you paid to a household employee.
- You are not a debtor in a Chapter 11 bankruptcy case filed after October 16, 2005.
- You do not claim any adjustments to income, such as a deduction for IRA contributions, a student loan interest deduction, an educator expenses deduction, or a tuition and fees deduction.
- You do not claim any credits other than the earned income credit.

1040A

If you did not meet these criteria for using the 1040EZ form, you could use Form 1040A if:

- Your income is only from wages, salaries, tips, taxable scholarships and fellowship grants, interest, or ordinary dividends, capital gain distributions, pensions, annuities, IRAs, unemployment compensation, taxable Social Security or railroad retirement benefits, and Alaska Permanent Fund dividends.

- Your taxable income is less than $100,000.
- You do not itemize deductions.
- You did not have an alternative minimum tax adjustment on stock you acquired by exercising an incentive stock option.
- Your taxes are only from the Tax Table, the alternative minimum tax, recapture of an education credit, Form 8615, or the Qualified Dividends and Capital Gain Tax Worksheet.
- Your only adjustments to income are the IRA deduction, the student loan interest deduction, the educator expenses deduction, and the tuition and fees deduction.
- The only credits you are claiming are the credit for child and dependent care expenses, the earned income credit, the credit for the elderly or the disabled, education credits, the child tax credit, the additional child tax credit, and the retirement savings contribution credit.

Differences Between the Two

Both Form 1040EZ and Form 1040A have an income requirement of less than $100,000. The differences involve the declaration of tips earned and declared as part of your income and the issue of deductions for an IRA, specific educational expenses, and credits for care of others.

You must file Form 1040 if your situation includes *any* of the following conditions:

- Your taxable income is $100,000 or more.
- You have certain types of income, such as unreported tips; certain nontaxable distributions; self-employment earnings; or income received as a partner, a shareholder in an "S" Corporation, or a beneficiary of an estate or trust.
- You owe household employment taxes.

FACT

"S" corporations "elect to pass corporate income, losses, deductions and credit through to their shareholders for federal tax purposes." (*www.irs.gov/Businesses/Small-Businesses-&-Self-Employed/S-Corporations*). You itemize deductions or claim certain tax credits or adjustments to income.

Income in excess of $100,000 automatically puts you in this category, as does any other factor in the foregoing list.

Completing the W-4 Form

Among the many forms you fill out when you begin a new job is the W-4 form. This form indicates how much money you want to have withheld from your paycheck to pay your income tax. In theory, this form is designed so that when you file your tax return the following spring, the balance of the tax bill should be zero. That is, you have already paid your share of taxes, no more and no less. In reality, however, it rarely works out that way.

The W-4 Step-by-Step

	Personal Allowances Worksheet (Keep for your records.)		
A	Enter "1" for **yourself** if no one else can claim you as a dependent	A	
B	Enter "1" if: { • You are single and have only one job; or / • You are married, have only one job, and your spouse does not work; or / • Your wages from a second job or your spouse's wages (or the total of both) are $1,500 or less. } . . .	B	
C	Enter "1" for your **spouse**. But, you may choose to enter "-0-" if you are married and have either a working spouse or more than one job. (Entering "-0-" may help you avoid having too little tax withheld.)	C	
D	Enter number of **dependents** (other than your spouse or yourself) you will claim on your tax return	D	
E	Enter "1" if you will file as **head of household** on your tax return (see conditions under **Head of household** above) . .	E	
F	Enter "1" if you have at least $1,900 of **child or dependent care expenses** for which you plan to claim a credit (**Note.** Do **not** include child support payments. See Pub. 503, Child and Dependent Care Expenses, for details.)	F	
G	**Child Tax Credit** (including additional child tax credit). See Pub. 972, Child Tax Credit, for more information. • If your total income will be less than $65,000 ($95,000 if married), enter "2" for each eligible child; then **less** "1" if you have three to six eligible children or **less** "2" if you have seven or more eligible children. • If your total income will be between $65,000 and $84,000 ($95,000 and $119,000 if married), enter "1" for each eligible child . . .	G	
H	Add lines A through G and enter total here. (**Note.** This may be different from the number of exemptions you claim on your tax return.) ▶ H		
	For accuracy, complete all worksheets that apply.	• If you plan to **itemize** or **claim adjustments to income** and want to reduce your withholding, see the **Deductions and Adjustments Worksheet** on page 2. • If you are **single** and have more than one job or are **married** and you and your spouse both work and the combined earnings from all jobs exceed $40,000 ($10,000 if married), see the **Two-Earners/Multiple Jobs Worksheet** on page 2 to avoid having too little tax withheld. • If **neither** of the above situations applies, **stop here** and enter the number from line H on line 5 of Form W-4 below.	

------------- Separate here and give Form W-4 to your employer. Keep the top part for your records. -------------

Form **W-4** Department of the Treasury Internal Revenue Service	**Employee's Withholding Allowance Certificate** ▶ Whether you are entitled to claim a certain number of allowances or exemption from withholding is subject to review by the IRS. Your employer may be required to send a copy of this form to the IRS.	OMB No. 1545-0074 **2013**
1 Your first name and middle initial	Last name	2 Your social security number
Home address (number and street or rural route)	3 ☐ Single ☐ Married ☐ Married, but withhold at higher Single rate. **Note.** If married, but legally separated, or spouse is a nonresident alien, check the "Single" box.	
City or town, state, and ZIP code	4 If your last name differs from that shown on your social security card, check here. You must call 1-800-772-1213 for a replacement card. ▶ ☐	

5	Total number of allowances you are claiming (from line **H** above **or** from the applicable worksheet on page 2)	5	
6	Additional amount, if any, you want withheld from each paycheck	6	$
7	I claim exemption from withholding for 2013, and I certify that I meet **both** of the following conditions for exemption. • Last year I had a right to a refund of **all** federal income tax withheld because I had **no** tax liability, **and** • This year I expect a refund of **all** federal income tax withheld because I expect to have **no** tax liability. If you meet both conditions, write "Exempt" here . . . . ▶	7	

Under penalties of perjury, I declare that I have examined this certificate and, to the best of my knowledge and belief, it is true, correct, and complete.

Employee's signature
(This form is not valid unless you sign it.) ▶   Date ▶

8 Employer's name and address (Employer: Complete lines 8 and 10 only if sending to the IRS.)	9 Office code (optional)	10 Employer identification number (EIN)

For Privacy Act and Paperwork Reduction Act Notice, see page 2.   Cat. No. 10220Q   Form **W-4** (2013)

Line A: As directed, enter 1 for yourself if no one else can claim you as a dependent. (Enter 0 if someone else, such as your parents, will be claiming you as a dependent.)

Line B: Enter 1 if you meet one of the criteria listed on the form (you work one job and your spouse does not work; or your wages from a second job or your spouse's wages are less than or equal to $1,500). Otherwise, enter 0.

Line C: The fewer dependents you claim, the greater the amount of money that will be taken from your paycheck for taxes, and the less likely it is that you will owe money in the spring when you file your tax return.

Line D: Enter the appropriate number.

Line E: The directions for head of household read, "Generally, you can claim head of household filing status on your tax return only if you are unmarried and pay more than 50% of the costs of keeping up a home for yourself and your dependent(s) or other qualifying individuals. See Pub. 501, Exemptions, Standard Deduction, and Filing Information, for information." Do you make more than half of the cost of keeping up your home? Enter 1 if this applies to you; enter 0 otherwise.

Line F: Enter 1 if you meet the criteria for child or dependent care paid out; enter 0 otherwise.

Line G: If your income as a single person is less than $65,000 ($95,000 for a married couple), use the following table to determine the appropriate number to enter.

Number of Children	1	2	3	4	5	6	7	8
Enter	2	4	5	7	9	11	12	14

Line H: Add the entries for lines A–G.

FACT

Line 6 on the W-4 form allows for you to have additional money taken from your check for tax purposes. This might be a reasonable option for someone who is unsure if the amount being withheld is less than the tax liability for the year.

Having done all of that, you will reach a final section that reads: "For accuracy, complete all worksheets that apply." This statement is adjacent to two statements about calculations you might want to make to ensure that you are paying the proper amount in taxes.

Form W-4 (2013)		Page **2**
Deductions and Adjustments Worksheet		

Note. Use this worksheet *only* if you plan to itemize deductions or claim certain credits or adjustments to income.

1	Enter an estimate of your 2013 itemized deductions. These include qualifying home mortgage interest, charitable contributions, state and local taxes, medical expenses in excess of 10% (7.5% if either you or your spouse was born before January 2, 1949) of your income, and miscellaneous deductions. For 2013, you may have to reduce your itemized deductions if your income is over $300,000 and you are married filing jointly or are a qualifying widow(er); $275,000 if you are head of household; $250,000 if you are single and not head of household or a qualifying widow(er); or $150,000 if you are married filing separately. See Pub. 505 for details . . .	**1**	$
2	Enter: { $12,200 if married filing jointly or qualifying widow(er) $8,950 if head of household $6,100 if single or married filing separately }	**2**	$
3	**Subtract** line 2 from line 1. If zero or less, enter "-0-"	**3**	$
4	Enter an estimate of your 2013 adjustments to income and any additional standard deduction (see Pub. 505)	**4**	$
5	**Add** lines 3 and 4 and enter the total. (Include any amount for credits from the *Converting Credits to Withholding Allowances for 2013 Form W-4* worksheet in Pub. 505.)	**5**	$
6	Enter an estimate of your 2013 nonwage income (such as dividends or interest)	**6**	$
7	**Subtract** line 6 from line 5. If zero or less, enter "-0-"	**7**	$
8	**Divide** the amount on line 7 by $3,900 and enter the result here. Drop any fraction	**8**	
9	Enter the number from the **Personal Allowances Worksheet**, line H, page 1	**9**	
10	**Add** lines 8 and 9 and enter the total here. If you plan to use the **Two-Earners/Multiple Jobs Worksheet**, also enter this total on line 1 below. Otherwise, **stop here** and enter this total on Form W-4, line 5, page 1	**10**	

FACT

You may benefit from itemizing your deductions on Form 1040, Schedule A, if you cannot use the standard deduction, had large uninsured medical and dental expenses, paid interest or taxes on your home, had large unreimbursed employee business expenses, had large uninsured casualty or theft losses, or made large charitable contributions.

If you plan on itemizing your deductions, you should complete the Deductions and Adjustments Worksheet. Line 4 of this table refers you to IRS Publication 505, which can be viewed at *www.irs.gov/pub/irs-pdf/ p505.pdf.*

Two-Earners/Multiple Jobs Worksheet (See *Two earners or multiple jobs* on page 1.)

Note. Use this worksheet *only* if the instructions under line H on page 1 direct you here.

1	Enter the number from line H, page 1 (or from line 10 above if you used the **Deductions and Adjustments Worksheet**)	1	
2	Find the number in **Table 1** below that applies to the **LOWEST** paying job and enter it here. **However,** if you are married filing jointly and wages from the highest paying job are $65,000 or less, do not enter more than "3"	2	
3	If line 1 is **more than or equal to** line 2, subtract line 2 from line 1. Enter the result here (if zero, enter "-0-") and on Form W-4, line 5, page 1. **Do not** use the rest of this worksheet	3	

Note. If line 1 is **less than** line 2, enter "-0-" on Form W-4, line 5, page 1. Complete lines 4 through 9 below to figure the additional withholding amount necessary to avoid a year-end tax bill.

4	Enter the number from line 2 of this worksheet 4		
5	Enter the number from line 1 of this worksheet 5		
6	**Subtract** line 5 from line 4	6	
7	Find the amount in **Table 2** below that applies to the **HIGHEST** paying job and enter it here	7 $	
8	**Multiply** line 7 by line 6 and enter the result here. This is the additional annual withholding needed . .	8 $	
9	Divide line 8 by the number of pay periods remaining in 2013. For example, divide by 25 if you are paid every two weeks and you complete this form on a date in January when there are 25 pay periods remaining in 2013. Enter the result here and on Form W-4, line 6, page 1. This is the additional amount to be withheld from each paycheck	9 $	

Table 1

Married Filing Jointly		All Others	
If wages from **LOWEST** paying job are—	Enter on line 2 above	If wages from **LOWEST** paying job are—	Enter on line 2 above
$0 - $5,000	0	$0 - $8,000	0
5,001 - 13,000	1	8,001 - 16,000	1
13,001 - 24,000	2	16,001 - 25,000	2
24,001 - 26,000	3	25,001 - 30,000	3
26,001 - 30,000	4	30,001 - 40,000	4
30,001 - 42,000	5	40,001 - 50,000	5
42,001 - 48,000	6	50,001 - 70,000	6
48,001 - 55,000	7	70,001 - 80,000	7
55,001 - 65,000	8	80,001 - 95,000	8
65,001 - 75,000	9	95,001 - 120,000	9
75,001 - 85,000	10	120,001 and over	10
85,001 - 97,000	11		
97,001 - 110,000	12		
110,001 - 120,000	13		
120,001 - 135,000	14		
135,001 and over	15		

Table 2

Married Filing Jointly		All Others	
If wages from **HIGHEST** paying job are—	Enter on line 7 above	If wages from **HIGHEST** paying job are—	Enter on line 7 above
$0 - $72,000	$590	$0 - $37,000	$590
72,001 - 130,000	980	37,001 - 80,000	980
130,001 - 200,000	1,090	80,001 - 175,000	1,090
200,001 - 345,000	1,290	175,001 - 385,000	1,290
345,001 - 385,000	1,370	385,001 and over	1,540
385,001 and over	1,540		

If you have more than one job and have an income in excess of $40,000, or if you and your spouse both work and have an income in excess of $10,000, complete the Two-Earners/Multiple Jobs Worksheet of the form.

FACT

Use Form 1040-ES to file estimated taxes for work done in which no taxes are deducted. This is true for independent workers, such as earnings from self-employment, interest, dividends, rents, and alimony. The government expects to be paid as money is earned and not wait until the end of the year. Waiting too long to pay the estimated tax can result in a penalty from the IRS.

Reading the W-2 Form

The W-2 Form is a record of the income earned, taxes paid, and (if applicable) contributions made to a retirement plan. Box a of the form shows the employee Social Security number, and Box b contains the employer's

ID number. Box c contains the employer's name and address, and box e contains the employee's name and address. Box d is where the employer inserts its control number if it uses one; many do not.

Box 1 (as opposed to a lettered box) reports the amount of wages, tips, and other compensation for the year. Box 2 records the federal tax withheld during the year. Boxes 3 and 5 contain the Social Security and Medicare wages and tips. If this number is different from Box 1, that is because it is the sum of the wages, etc., from Box 1 and the amount of pre-tax money you had taken out for your retirement plan. Boxes 4 and 6 report the amount of taxes paid for Social Security and Medicare, respectively. Box 15 has the two-letter abbreviation for your state, and line 17 has the amount of state income tax you paid.

Filling Out Form 1040EZ

You can download Form 1040EZ and the instructions for completing this form at *www.irs.gov*. Complete your identification information—name, address, and Social Security number—and your spouse's information if you are filing jointly. If you so choose, check the box donating $3 to the Presidential Election Campaign.

- Line 1—Wages, salaries, and tips. This number appears in Box 1 on your W-2 form.
- Line 2—Taxable interest received during the year. (A note on the back of Form 1040EZ states that you must include all taxable interest, including interest from banks, savings and loan institutions, credit unions, and any other financial institution, even if that institution did not send you the proper form, Form 1099-NT, with which to report the interest.) Enter -0- if there is no taxable interest to declare.
- Line 3—Unemployment insurance and Alaska Permanent Fund dividends received during the year, if applicable. The directions for completing Form 1040EZ explain how to complete this line if you, or your spouse in the case of a joint return, received too much money or repaid money toward unemployment insurance (money returned to the government because you or your spouse received more money than you were entitled to receive). Enter -0- if this does not apply to you.

Form 1040EZ

Department of the Treasury—Internal Revenue Service

Income Tax Return for Single and Joint Filers With No Dependents (99) **2012**

OMB No. 1545-0074

Your first name and initial | Last name | Your social security number

If a joint return, spouse's first name and initial | Last name | Spouse's social security number

Home address (number and street). If you have a P.O. box, see instructions. | Apt. no.

▲ Make sure the SSN(s) above are correct.

City, town or post office, state, and ZIP code. If you have a foreign address, also complete spaces below (see instructions).

Presidential Election Campaign
Check here if you, or your spouse if filing jointly, want $3 to go to this fund. Checking a box below will not change your tax or refund. ☐ You ☐ Spouse

Foreign country name | Foreign province/state/county | Foreign postal code

Income

Attach Form(s) W-2 here.

Enclose, but do not attach, any payment.

1   Wages, salaries, and tips. This should be shown in box 1 of your Form(s) W-2. Attach your Form(s) W-2. .... **1**

2   Taxable interest. If the total is over $1,500, you cannot use Form 1040EZ. **2**

3   Unemployment compensation and Alaska Permanent Fund dividends (see instructions). **3**

4   Add lines 1, 2, and 3. This is your **adjusted gross income.** **4**

5   If someone can claim you (or your spouse if a joint return) as a dependent, check the applicable box(es) below and enter the amount from the worksheet on back.
☐ **You**   ☐ **Spouse**
If no one can claim you (or your spouse if a joint return), enter $9,750 if **single**; $19,500 if **married filing jointly.** See back for explanation. **5**

6   Subtract line 5 from line 4. If line 5 is larger than line 4, enter -0-. This is your **taxable income.** ▶ **6**

Payments, Credits, and Tax

7   Federal income tax withheld from Form(s) W-2 and 1099. .... **7**

8a   **Earned income credit (EIC)** (see instructions). **8a**

b   Nontaxable combat pay election.   8b

9   Add lines 7 and 8a. These are your **total payments and credits.** ▶ **9**

10   **Tax.** Use the amount on **line 6 above** to find your tax in the tax table in the instructions. Then, enter the tax from the table on this line. **10**

Refund

Have it directly deposited! See instructions and fill in 11b, 11c, and 11d or Form 8888.

11a   If line 9 is larger than line 10, subtract line 10 from line 9. This is your **refund.** If Form 8888 is attached, check here ▶ ☐   **11a**

▶ b   Routing number [   ]   ▶ c Type: ☐ Checking ☐ Savings

▶ d   Account number [   ]

Amount You Owe

12   If line 10 is larger than line 9, subtract line 9 from line 10. This is the **amount you owe.** For details on how to pay, see instructions. ▶ **12**

Third Party Designee

Do you want to allow another person to discuss this return with the IRS (see instructions)? ☐ **Yes.** Complete below. ☐ **No**

Designee's name ▶ | Phone no. ▶ | Personal identification number (PIN) ▶ [   ]

Sign Here

Joint return? See instructions.

Keep a copy for your records.

Under penalties of perjury, I declare that I have examined this return and, to the best of my knowledge and belief, it is true, correct, and accurately lists all amounts and sources of income I received during the tax year. Declaration of preparer (other than the taxpayer) is based on all information of which the preparer has any knowledge.

Your signature | Date | Your occupation | Daytime phone number

Spouse's signature. If a joint return, **both** must sign. | Date | Spouse's occupation | If the IRS sent you an Identity Protection PIN, enter it here (see inst.) [   ]

Paid Preparer Use Only

Print/Type preparer's name | Preparer's signature | Date | Check ☐ if self-employed | PTIN

Firm's name ▶ | | Firm's EIN ▶

Firm's address ▶ | | Phone no.

For Disclosure, Privacy Act, and Paperwork Reduction Act Notice, see instructions.   Cat. No. 11329W   Form **1040EZ** (2012)

Use this form if

• Your filing status is single or married filing jointly. If you are not sure about your filing status, see instructions.
• You (and your spouse if married filing jointly) were under age 65 and not blind at the end of 2012. If you were born on January 1, 1948, you are considered to be age 65 at the end of 2012.

• You do not claim any dependents. For information on dependents, see Pub. 501.
• Your taxable income (line 6) is less than $100,000.
• You do not claim any adjustments to income. For information on adjustments to income, use the TeleTax topics listed under *Adjustments to Income* at *www.irs.gov/taxtopics* (see instructions).
• The only tax credit you can claim is the earned income credit (EIC). The credit may give you a refund even if you do not owe any tax. You do not need a qualifying child to claim the EIC. For information on credits, use the TeleTax topics listed under *Tax Credits* at *www.irs.gov/taxtopics* (see instructions). If you received a Form 1098-T or paid higher education expenses, you may be eligible for a tax credit or deduction that you must claim on Form 1040A or Form 1040. For more information on tax benefits for education, see Pub. 970.

• You had only wages, salaries, tips, taxable scholarship or fellowship grants, unemployment compensation, or Alaska Permanent Fund dividends, and your taxable interest was not over $1,500. But if you earned tips, including allocated tips, that are not included in box 5 and box 7 of your Form W-2, you may not be able to use Form 1040EZ (see instructions). If you are planning to use Form 1040EZ for a child who received Alaska Permanent Fund dividends, see instructions.

Filling in your return

If you received a scholarship or fellowship grant or tax-exempt interest income, such as on municipal bonds, see the instructions before filling in the form. Also, see the instructions if you received a Form 1099-INT showing federal income tax withheld or if federal income tax was withheld from your unemployment compensation or Alaska Permanent Fund dividends.

For tips on how to avoid common mistakes, see instructions.

Remember, you must report all wages, salaries, and tips even if you do not get a Form W-2 from your employer. You must also report all your taxable interest, including interest from banks, savings and loans, credit unions, etc., even if you do not get a Form 1099-INT.

Worksheet for Line 5 — Dependents Who Checked One or Both Boxes

Use this worksheet to figure the amount to enter on line 5 if someone can claim you (or your spouse if married filing jointly) as a dependent, even if that person chooses not to do so. To find out if someone can claim you as a dependent, see Pub. 501.

A. Amount, if any, from line 1 on front . . . . . .
          +   300.00 Enter total ▶ **A.** _____

B. Minimum standard deduction . . . . **B.** _____950.00_____

C. Enter the **larger** of line A or line B here . . . . **C.** _____

D. Maximum standard deduction. If **single,** enter $5,950; if **married filing jointly,** enter $11,900 . **D.** _____

E. Enter the **smaller** of line C or line D here. This is your standard deduction . . . . **E.** _____

F. Exemption amount.
    • If single, enter -0-.
    • If married filing jointly and —
        —both you and your spouse can be claimed as dependents, enter -0-.
        —only one of you can be claimed as a dependent, enter $3,800.   **F.** _____

G. Add lines E and F. Enter the total here and on line 5 on the front . . . . **G.** _____

(keep a copy for your records)

If you did not check any boxes on line 5, enter on line 5 the amount shown below that applies to you.
• Single, enter $9,750. This is the total of your standard deduction ($5,950) and your exemption ($3,800).
• Married filing jointly, enter $19,500. This is the total of your standard deduction ($11,900), your exemption ($3,800), and your spouse's exemption ($3,800).

Mailing Return

Mail your return by **April 15, 2013.** Mail it to the address shown on the last page of the instructions.

Form **1040EZ** (2012)

- Line 4—As directed, add lines 1, 2, and 3 to determine your gross income for the year.
- Line 5—Check one or both of the boxes if someone else can claim you and/or your spouse as a dependent. If you check either one, complete the worksheet for line 5.

Worksheet for Line 5 — Dependents Who Checked One or Both Boxes	Use this worksheet to figure the amount to enter on line 5 if someone can claim you (or your spouse if married filing jointly) as a dependent, even if that person chooses not to do so. To find out if someone can claim you as a dependent, see Pub. 501.

A. Amount, if any, from line 1 on front . . . .

+ _____300.00__ Enter total ▶   **A.** _____

B. Minimum standard deduction . . . . **B.** ____950.00____

C. Enter the **larger** of line A or line B here . . . . **C.** _____

D. Maximum standard deduction. If **single**, enter $5,950; if **married filing jointly**, enter $11,900   .   **D.** _____

E. Enter the **smaller** of line C or line D here. This is your standard deduction . . . .   **E.** _____

F. Exemption amount.
  • If single, enter -0-.
  • If married filing jointly and —
    —both you and your spouse can be claimed as dependents, enter -0-.
    —only one of you can be claimed as a dependent, enter $3,800.   **F.** _____

G. Add lines E and F. Enter the total here and on line 5 on the front   . . . .   **G.** _____

(keep a copy for your records)

If you did not check any boxes on line 5, enter on line 5 the amount shown below that applies to you.
• Single, enter $9,750. This is the total of your standard deduction ($5,950) and your exemption ($3,800).
• Married filing jointly, enter $19,500. This is the total of your standard deduction ($11,900), your exemption ($3,800), and your spouse's exemption ($3,800).

- Line 6—As directed, subtract line 5 from line 4. If the answer is a negative number, enter -0-.
- Line 7—Record the federal income tax withheld. You will find this in Box 2 on your W-2 form.
- Line 8—Earned income credit. The directions for completing this line involve answering a series of questions that can be found in the instruction manual for filing Form 1040EZ.
- Line 9—As directed, add lines 7 and 8a.
- Line 10—Use the amount on line 6 to determine the amount of tax you owe for the year. (The tax table is located near the end of the document.)
- Line 11—If line 9 is larger than line 10, you paid too much in taxes and are entitled to a refund. Subtract line 10 from line 9 to determine the amount of the refund. You can include routing information on line 11 if you would like an electronic payment of your refund.
- Line 12—If line 10 is larger than line 9, you owe money to the government. Subtract line 9 from line 10 to determine the amount you owe.

- Third Party Designee—You may designate a third person to deal with the IRS regarding any questions about your tax return. This includes the IRS calling the third party or the third party calling the IRS. If you check Yes, the PIN (personal identification number) is chosen by the third party and is used as a means of identification when that third party talks with someone from the IRS.

Be sure to sign and date your return, indicate your occupation, and supply a daytime telephone number where you can be reached (should there be any questions).

ESSENTIAL

You may want to have a tax professional examine your 1040EZ Form if this is the first time you are filing one on your own, if that would give you peace of mind. If you are required to file Form 1040 or 1040A, you may want to use a tax professional. You get the benefit of their expertise and they will represent you should the government have questions about your tax return.

Examples of Filing Taxes

Example: Mairan, filing as a single person, has her W-2 Form. According to the form, her wages, tips, and other compensation for the year totaled $39,816.54. The amount of federal tax withheld was $5,972.48. She had $25.10 in taxable interest for the year. She cannot be claimed by anyone else as a dependent, and she does not qualify for any earned income credit, nor has she qualified for unemployment insurance or the Alaska Permanent Fund dividend. She uses Form 1040EZ to file her tax return.

Mairan enters $39,816.54 on line 1, $25.10 on line 2, and -0- on line 3. She adds these numbers together to get her adjustable gross income, $39,841.64, and writes this on line 4. Since no else can claim her as a dependent and she is filing as a single person, she enters $9,750 (the deduction allowed for a single person with no dependents) on line 5. She subtracts line 5 from line 4 to get a taxable income of $30,091.64 for line 6.

She records $5,972.48 on line 7 as the amount of federal tax withheld during the year; -0- on line 8, because she is not eligible for any earned tax credits; and $5,972.48 on line 9 for the total payment of federal taxes and credits for the year.

The Tax Table for the 2012 tax year is found on page 31 of the instructions manual (*www.irs.gov*). Mairan's taxable income, listed on line 6 of Form 1040EZ, is $30,091.64. She goes to page 34, where she can see the taxable income for the numbers in the $30,000 range. Her taxable income is between $30,050 and $30,100. As a single person, her tax bill is $4,076. She records this number on line 10.

Because line 9 is greater than line 10, Mairan is entitled to a refund of $1,896.48 ($5,972.48 − $4,076). She can choose to have her refund deposited electronically into her bank account, or the government can send her a check for the amount due.

How is Mairan's situation different if she checks Yes on line 5, to indicate that she can be declared as a dependent on someone else's income tax return? In this case, using the worksheet on page 2 of the tax form, she enters $39,816.54 in the box for statement A and then adds 300. She enters the total, $40,116.54, on line A. She also enters $40,116.54 on line C. She puts $5,950 on line D and again on line E. Line F is -0- and line G is $5,950. This goes on line 5 of Form 1040EZ.

She subtracts line 5 from line 4 to get a taxable income of $33,866.54 for line 6.

She records $5,972.48 on line 7 as the amount of federal tax withheld during the year; -0- on line 8, because she is not eligible for any earned tax credits; and $5,972.48 on line 9 for the total payment of federal taxes and credits for the year.

Mairan's taxable income, listed on line 6 of Form 1040EZ, is $33,866.54. She goes to page 34, where she can see the taxable income for the numbers in the $30,000 range. Her gross taxable income is between $33,850 and $33,900. As a single person, her tax bill is $4,646. She records this number on line 10.

Because line 9 is greater than line 10, Mairan is entitled to a refund of $1,326.48 ($5,972.48 − $4,646).

How do you use math when you file your taxes?
Consider the impact of IRAs on reducing the amount of taxable income you need to report. What else did you add to your list of when you use math?

Exercises

Complete Form 1040EZ for each of these returns.

1. Kieran, filing as a single person, has his W-2 Form. According to the form, his wages, tips, and other compensation for the year totaled $49,167.49. The amount of federal tax withheld was $8,358.47. He had $51.15 in taxable interest for the year. He cannot be claimed by anyone else as a dependent, and he does not qualify for any earned income credit, nor has he qualified for unemployment insurance or the Alaska Permanent Fund dividend.

2. Allie, filing as a single person, has her W-2 Form. According to the form, her wages, tips, and other compensation for the year totaled $89,978.54. The amount of federal tax withheld was $15,296.35. She had $125.10 in taxable interest for the year. She cannot be claimed by anyone else as a dependent, and she does not qualify for any earned income credit, nor has she qualified for unemployment insurance or the Alaska Permanent Fund dividend.

CHAPTER 15

Probability and Statistics

Desktop computers have had a tremendous impact on the world, especially by boosting people's ability to collect data. Once data are collected, however, they have to be analyzed—a process so important that the phrase *data-driven decisions* (or some variation of it, such as *data-based marketing*) is used in every industry. Much of the time, to make an informed decision about how to interpret data, you have to consider statistics. Statistics has two branches. *Descriptive statistics* yields information about the center and spread of data. (You will learn more about these two concepts in the discussion to come.) *Inferential statistics*, the application of probability theory, is a tool used to make data-driven decisions. The most basic elements of probability and statistics are learning how to count and determining where the center of your data is located.

Simple Events—Flipping a Coin

In the language of probability, a simple event is an event in which one outcome is examined. For instance, flipping a coin, rolling a die, spinning a spinner, drawing a ticket from a hat for a 50-50 drawing, and playing a number at a roulette table are all simple events. (In a 50-50 drawing, the winner takes half of the pot, and the sponsoring organization takes the other half. Such drawings are popular as fundraisers for charities.) Before you continue learning about simple events—and probability in general—there is some language that you should be familiar with.

FACT

The study of probability did not begin until the mid 17th century when Blaise Pascal and Pierre de Fermat were asked to examine a problem about gambling.

The Language of Probability

- *Probability* is the relative frequency with which an event will occur. In simpler language, probability is the ratio of the number of ways a particular event can occur to the number of ways all possible events can occur.
- A *fair event* is one in which all possible events have the same chance of occurring, whereas in a *biased event*, some outcomes are more likely than others. (You may have heard of "loaded dice" being used to cheat in gambling. Loaded dice have been altered so that certain numbers are more likely to occur than others.)

Flip a fair coin. The probability that it will show Heads is $\frac{1}{2}$.

Roll a fair die. The probability that the die will show a 5 is $\frac{1}{6}$.

Draw a ticket from a hat for a 50-50 drawing. The probability that you will win depends on two factors: how many tickets were sold and how many you bought. If 250 tickets were sold for the drawing and you bought 1 ticket, the probability that you will win is $\frac{1}{250}$. However, if you bought 25 tickets, your chance of winning increases to $\frac{25}{250}$, or $\frac{1}{10}$.

Compound Events—Rolling the Dice

Computing probabilities for events that require more than one outcome to occur (such as getting the coin to land showing Heads and rolling a 5 when you toss a die) is more complicated. The Fundamental Theorem of Counting (fundamental in the sense of "basic" and "essential") helps to determine the number of possible outcomes for a compound situation. The basic statement of the Fundamental Theorem of Counting is this: When an experiment is performed in a series of steps, the number of possible outcomes for the experiment as a whole is equal to the product of the numbers of possible outcomes for all the steps in the series taken individually. (It's less complicated than it sounds! Read on.)

For example, the number of possible outcomes when you toss a coin *and* roll a die is equal to the number of possible outcomes when you toss a coin (which is 2) times the number of possible outcomes when you roll a die (which is 6). Thus the number of outcomes when you toss a coin *and* roll a die is $2 \times 6 = 12$. The number of "successful outcomes" for the experiment is the number of ways you can get the coin to land showing Heads (which is 1) multiplied by the number of ways you can roll a 5 (which is 1), for a product of $1 \times 1 = 1$. Thus the probability of getting Heads and a 5 in this experiment is $\frac{1}{12}$.

When you roll two dice, there are a total of 36 outcomes (6 × 6). Much of the time, when people are rolling two dice, they need to add the face values of the dice.

	1	2	3	4	5	6
1	2	3	4	5	6	7
2	3	4	5	6	7	8
3	4	5	6	7	8	9
4	5	6	7	8	9	10
5	6	7	8	9	10	11
6	7	8	9	10	11	12

Take the time to look at the table. There is one outcome where the sum is 2 and one outcome where the sum is 12. There are two outcomes where the sum is 3 and two outcomes where the sum is 11; there are three outcomes where the sum is 4 and three outcomes where the sum is 10; there are four outcomes where the sum is 5 and four outcomes where the sum is 9; there are five outcomes where the sum is 6 and five outcomes where the sum is 8; and there are six outcomes where the sum is 7.

ESSENTIAL

Graphic representations of the outcomes for probability experiments are often used to clarify the set of possible outcomes (the sample space). The two most-popular graphical representations are the two-way table and the tree diagram.

What is the probability that when two fair dice are rolled, the sum will be 8? Of the 36 outcomes, there are 5 ways to get an 8, so the probability of getting an 8 is $\frac{5}{36}$. What is the probability that the sum will be a 6 or a 10? There are 8 outcomes that yield either a 6 or a 10, so the probability will be $\frac{10}{36}$.

For Example . . .

The restaurants in many communities take a week out of each year and designate it "Restaurant Week." During this time, each restaurant offers a limited selection of three-course meals at a fixed price. In this context, a meal is defined as 1 appetizer, 1 entrée, and 1 dessert. The reduced menu consists of 2 appetizers, 4 entrées, and 2 desserts, and the diner can choose one of each. Thus there are $2 \times 4 \times 2 = 16$ different meals available. Can you calculate how many different meals are available if the menu is changed to offer a choice among 3 appetizers, 5 entrées, and 2 desserts? That's right, there are now $3 \times 5 \times 2 = 30$ different meals available. Bon appétit!

Automated teller machines (ATMs) require users to enter a 4-digit personal identification number (PIN) when they use the account. With the 10 digits available on the keypad, there are $10 \times 10 \times 10 \times 10 = 10,000$ different PINs available. (Note that, given the amount of activity that occurs at an ATM in a densely populated area, there is a high probability that more than two people will have the same PIN.)

A survey is taken of the voters in a county about their reactions to the county's budget. Information about the gender and political affiliation of the voters surveyed is displayed in a table like this:

	Democrats	Republicans	Independents	Total
Women	183	172	45	400
Men	178	181	67	426
Total	361	353	112	26

If a voter is selected at random, what is the probability that this voter is a woman *and* is a Democrat? The table shows that of the 826 people surveyed, 183 are women who identify themselves as Democrats, so the probability of the outcome is $\frac{183}{826}$.

If a voter is selected at random, what is the probability that the voter is a male *or* has a Republican affiliation? The table shows that there are 426 males and 353 Republicans. However, there are 181 people who are counted in both groups and should be counted only once. The number of people who are men or Republicans is equal to the number of men plus the number of Republicans minus the number who are both Republicans and men. As a "formula," Men or Republicans = Men + Republicans − (Men and Republicans). There are $426 + 353 − 181 = 598$ voters who are men or Republicans, so the probability of the outcome is $\frac{598}{826}$.

ESSENTIAL

In the language of probability, *and* is an indication to multiply, and *or* is an indication to add.

Look back at the table representing the outcomes for rolling two dice. What is the probability that the sum of the faces is 8 or the two dice show the same number? (Rolling the same number on both dice is known as rolling doubles.) The number of outcomes that yield a sum of 8 is 5; and of the 6 outcomes for doubles, 1 of the doubles (4 and 4) gives a sum of 8. The probability of getting a sum of 8 or doubles is therefore $\frac{5 + 6 − 1}{36} = \frac{10}{36}$.

A standard deck of bridge cards consists of 52 cards. There are 26 red cards and 26 black cards. The red cards are divided into the suits Hearts and Diamonds, and the black cards are divided into Spades and Clubs. Each suit has 13 cards: Ace, 2, 3, 4, 5, 6, 7, 8, 9, 10, Jack, Queen, and King. The Jacks, Queens, and Kings have the images of people on them and are called picture cards, or sometimes "face cards." (This means that there are 13 cards in each suit; there are 4 cards with each face value; and there are 26 black cards and 26 red cards. Say a card is randomly selected from a well-shuffled deck (you want to make sure that the outcome is completely random and not predetermined, or biased, by the placing of the cards). What is the probability that the card selected is a Heart or a picture card? There are 13 Hearts and 12 picture cards, but 3 of the picture cards are Hearts. The probability that the selected card is a Heart or a picture card is $\frac{13 + 12 − 3}{52} = \frac{22}{52}$.

What Are the Odds?

Roll a fair die. Take note of the face value. Roll the die a second time. What is the probability that both rolls show a 3? The probability that the first roll is a 3 is $\frac{1}{6}$, and the probability that the second roll is a 3 is also $\frac{1}{6}$. The probability that the first roll and the second roll show a 3 is $\frac{1}{6} \times \frac{1}{6} = \frac{1}{36}$. (For the record, because the outcome for each of the numbers 1 through 6 on a die is $\frac{1}{6}$, the answer for all possible outcomes for a roll of two dice is $\frac{1}{36}$.)

Flip a fair coin three times. The probability of getting 3 Heads is $\frac{1}{8}$ ($\frac{1}{2} \times \frac{1}{2} \times \frac{1}{2}$). What is the probability of getting 2 Heads in the three tosses? (In the language of probability, this means getting exactly 2 Heads, not 2 or more.) As in the example of two dice in the last paragraph, the probability of any specified three-coin outcome will be $\frac{1}{8}$. That is, the probability of HHH (3 Heads), that of HTH (Heads Tails Heads), and that of TTH (Tails Tails Heads) are all the same: $\frac{1}{8}$. Which outcomes from all the possible outcomes contain 2 Heads? Well, there is HHT, and there are also HTH and THH. There are a total of three different ways in which the outcomes can contain 2 Heads, so $3 \frac{1}{8} = \frac{3}{8}$.

FACT

The Thoroughbred Orb was the 2013 favorite to win the Kentucky Derby with the odds listed at 7:2. What do these numbers mean? In the language of betting, this means that in the next 9 races, Orb will lose 7 and win 2. Bettors' odds are listed as the ratio of the number of ways the event will not happen to the number of ways the event will happen.

Playing with Cards

Three cards are to be drawn from a well-shuffled bridge deck of cards. The first card will be drawn, its suit will be noted, and then the card will be returned to the deck. The deck will be shuffled before the second card is drawn, noted, and returned to the deck, and this process will be repeated for the third card. What is the probability that 3 Spades will be drawn? For each drawing, the probability of drawing a Spade is $\frac{1}{4}$, so the probability of getting a Spade *and* a Spade *and* a Spade is $\frac{1}{4} \times \frac{1}{4} \times \frac{1}{4} = \frac{1}{64}$. What is the probability that there will be 2 Spades in the drawing of the three cards? In the case of flipping a coin, the outcome is either Heads (H) or Tails (T) for each flip. In the case of picking a card, the outcome of interest is a Spade (S) or some other (O) suit (whether that other suit is Diamonds, Clubs, or Hearts is not important—it can't be a Spade). The outcomes for the three cards are SSO, SOS, or OSS. Unlike the case of the coin, where each outcome H or T has the probability of $\frac{1}{2}$, the probability of drawing the "other" card is $\frac{3}{4}$. Therefore, the outcome SSO has the probability of $\frac{1}{4} \times \frac{1}{4} \times \frac{3}{4} = \frac{3}{64}$, the probability of SOS is $\frac{1}{4} \times \frac{3}{4} \times \frac{1}{4} = \frac{3}{64}$, and the probability of OSS is $\frac{3}{4} \times \frac{1}{4} \times \frac{1}{4} = \frac{3}{64}$. The probability of drawing exactly 2 Spades is therefore $3 \frac{3}{64} = \frac{9}{64}$.

Here's the same question but with one change: Three cards are to be drawn from a well-shuffled bridge deck of cards. The first card will be drawn, its suit will be noted, and then the card will *not* be returned to the deck. The deck will be shuffled before the second card is drawn, noted, and not returned to the deck, and this process will be repeated for the third card. What is the probability that 3 Spades will be drawn? The probability that the first card is a Spade is $\frac{13}{52} = \frac{1}{4}$. Assuming that the first card picked is a Spade, the deck will then contain 51 cards, 12 of which are Spades.

The probability that the second card selected is a Spade is $\frac{12}{51}$, and the probability the third card selected is a Spade is $\frac{11}{50}$. The probability of getting three Spades is thus $\frac{13}{52} \times \frac{12}{51} \times \frac{11}{50} = \frac{1716}{132600} = \frac{11}{850}$.

What is the probability that there will be 2 Spades in the drawing of the three cards? The SSO outcome has the probability of $\frac{13}{52} \times \frac{12}{51} \times \frac{39}{50} = \frac{6084}{132600} = \frac{39}{850}$.

The SOS outcome has the probability of $\frac{13}{52} \times \frac{39}{51} \times \frac{12}{50} = \frac{6084}{132600} = \frac{39}{850}$. Do you see that the numerators for the probability of the OSS outcome will 39, 13, and 12, in that order, and that the probability will also be $\frac{6084}{132600} = \frac{39}{850}$.

The probability of getting 2 Spades is $3 \frac{6084}{132600} = \frac{18252}{132600} = \frac{117}{850}$.

Will You Win the Lottery?

Winning the big prize for the Powerball lottery game requires that you correctly guess the 5 white numbers in the drawing (out of a possible 59) and the 1 red number (out of a possible 35). What is the probability of winning the big prize? The probability that you have the first white ball drawn is $\frac{5}{59}$, that you have the second white ball is $\frac{4}{58}$, that you have the third white ball is $\frac{3}{57}$, that you have the fourth white ball is $\frac{2}{56}$, that you have the fifth white ball is $\frac{1}{55}$, and that you have the red ball is $\frac{1}{35}$. The probability of winning the big prize is $\frac{5}{59} \times \frac{4}{58} \times \frac{3}{57} \times \frac{2}{56} \times \frac{1}{55} \times \frac{1}{35} = \frac{120}{21,026,821,200} = \frac{1}{175,223,510}$.

Where's the Center?—Averages

Imagine that you overhear the following conversation between Geraldo and Alice:

Alice: "I got my psychology test back today."
Geraldo: "How did you do?"
Alice: "I got 82%."
Geraldo: "Is that a good score?"

Geraldo's last question might seem odd at first. After all, Alice scored 82%, 18 percentage points away from a perfect paper. But consider two different scenarios that might occur when Alice responds to this question.

1. Scenario I—Alice: "The average grade on the test was 64%."
2. Scenario II—Alice: "The average grade on the test was 93%."

In the first scenario, Alice scored well above the average, and in the second case, she did not. The term *average* is used quantitatively and qualitatively to indicate a sense of, well, average. (You could use the word *typical* as an indication of what is usual, or you could use the word *normal*, but that has other meanings in the world of statistics.) In essence, the average is an indication of where the center of the group is and how values (both quantitative and qualitative) are compared against this center.

In a typical situation, most values are located near the center of the data, and the farther you move away from the center, the values appear less and less frequently. For example, the heights of 16-year-old males are taken in the high schools of Central County. The heights of the 1,000 students whose heights were measured are displayed in the histogram.

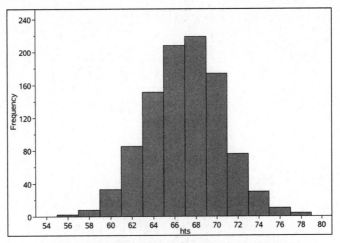

Heights of 16-Year-Old Students in Central County

Notice that most of the students have a height near the 66–68-inch bars on the histogram. There are fewer students whose heights are much larger or smaller than this central value. In fact, many people believe that data for most situations, from standardized test scores to salary distributions, fit the shape of the bell curve. (The bell curve is called a *normal distribution*, reinforcing the notion that data are typically distributed in this way. You can think of the name in this way, if the expectation that data will fit in this way, most values near the mean and less data as you go farther from the mean, you might say, "Well, that's normal" in that this is what you would expect while any other arrangement of the data would seem not quite right.)

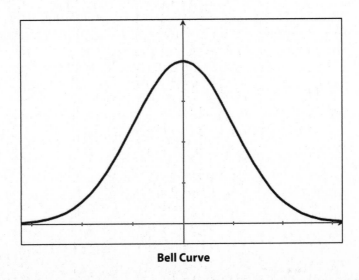

Bell Curve

As you can see from the distribution of major league baseball salaries on the beginning of opening day 2013 (Source: *www.usatoday.com/sports/ mlb/salaries/2013/all/team/all/*), this is not so.

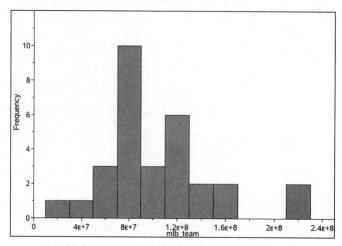

Major League Baseball Salaries per Team

Finding the Average

For most people, the word *average* indicates that a set of numbers is added together, and then the sum is divided by the number of pieces of data. For example, the average of 82, 87, 91, 93, and 84 is $\dfrac{82+87+91+93+84}{5} = \dfrac{437}{5} = 87.4$. This number, whose official name is the *mean*, is the predominant measure of center, and that is exactly how it is found. However, there are two other measures of center, the median and the mode, that are also called averages.

The *mode* (think "most") is that data value that occurs more often than any other value. For example, the mode for the data 10, 10, 10, 10, 13, 15, 17, 19, 23, 29, and 35 is 10. It is the number that occurs the most often (the data value with the highest frequency). Although it is a statistically valid measure of center, the mode is the least often used of the three.

The *median* is truly the center, because of the way it is computed. The data are arranged in increasing (or decreasing) order. If there are an odd number of data values, the median is the central number. If there are an even number of data points, the median is midway between the two middle values. For example, the median for the data 10, 10, 10, 10, 13, 15, 17, 19, 23, 29, and 35 is 15. The five data points 10, 10, 10, 10, and 13 are below it, and the five data points 17, 19, 23, 29, and 35 are above it. If another value, say 25, is added to the data set, then the data arranged in increasing order will be 10, 10, 10, 10, 13, 15, 17, 19, 23, 25, 29, and 35. The two numbers in the middle are 15 and 17, so the median will be 16 ($\frac{15+17}{2}$).

FACT

An Excel spreadsheet has commands called Average and Median that can be used to compute these measures of center.

Example: Find the mean and the median for these data:
20, 23, 43, 19, 59, 93, 90, 83, 74, 79, 29, 10, 31, 45, 59, 43, 23, 39

You can calculate the mean by adding the data values and dividing by the number of data points.

$$\frac{20 + 23 + 43 + 19 + 59 + 93 + 90 + 83 + 74 + 79 + 29 + 10 + 31 + 45 + 59 + 43 + 23 + 39}{18} = \frac{862}{18} = 47.889$$

To calculate the median, you start by arranging the data.
10, 19, 20, 23, 23, 29, 31, 39, 43, 43, 45, 59, 59, 74, 79, 83, 90, 93
There are 18 pieces of data. Find the ninth and tenth pieces of data, 43 and 43. The mean of these two numbers is 43, so the median for the data set is 43.

Frequency Table

When data values are repeated a large number of times, the data are recorded in a frequency table. The accompanying table shows the number of meals served in an evening at a restaurant over a period of time.

Dinners Served	Frequency
45	27
48	32
50	40
52	48
54	43
55	38

The table shows that the restaurant served 45 dinners 27 different times, 48 dinners 32 times, and so on. The sum of the Frequency column, 228, indicates the number of data points. The median for these data is the mean of the 114th and 115th pieces of data. The first three values (45, 48, and 50) make up $27 + 32 + 40 = 95$ pieces of data. The next data value, 52, makes up the next 48 data values, and this will include the 114th and 115th data values. The median for the number of dinners served in an evening for this restaurant is 52. The mean for these data is found by adding all 228 values—a lot of work. Fortunately, multiplication can be used to simplify the process. The number of dinners served is $45 \times 27 + 48 \times 32 + 50 \times 40 + 52 \times 48 + 54 \times 43 + 55 \times 38 = 11{,}659$. Divide this by 228 to get the mean, 51.1.

Using Excel

There are two ways to use Excel to find the mean. In each case, enter the data into adjacent columns.

The first way to do this is to go to cell C1 and type $= A1*B1$ and press Enter to multiply the values in cells A1 and B1. Highlight this cell and choose Edit Copy. Move the cursor to cell C2, click the mouse, and drag the mouse to cell C6 (drawing a dotted rectangle, or marquee, around the cells). Press Edit and Paste to paste the formulas to the remaining cells in column C.

	A	B	C
1	45	27	1215
2	48	32	1536
3	50	40	2000
4	52	48	2496
5	54	43	2322
6	55	38	2090
7			

Move the cursor to cell C7, and enter = Sum(C1:C6)/Sum(B1:B6) to get the answer 51.1.

The second method uses an Excel command that is the equivalent of entering the multiplication cell into C1, copying this formula into cells C2 through C6, and adding the entries in cells C1 through C6. It is call the SUMPRODUCT command. Move the cursor to cell B7, and type = SUMPRODUCT(A1:A6,B1:B6) there. Then type = B7/Sum(B1:B6) in cell B8.

	A	B
1	45	27
2	48	32
3	50	40
4	52	48
5	54	43
6	55	38
7	sumproduct(a1:a6,b1:b6)=	11659
8	b7/sum(b1:b6)	51.13596491

The mode for these data is easily found, because it is the data value with the highest frequency. In this case, the mode is 52.

A Word to the Wise

A word of caution is necessary here. Although the mean is the most popular measure of where the data center is, it is also the most susceptible to outliers (numbers completely out of line with the rest of the data). For example, consider the salaries of the 6 employees of a small business (numbers in thousands of dollars): 25, 27, 31, 32, 34, and 38. The mean for these data is $31.167 thousand, and the median is $31.5 thousand. But then a seventh employee is hired with a salary of $100,000. The data are now 25, 27, 31, 32, 34, 38, 100, and this changes the mean to $41 thousand and the median is $32 thousand. This one salary raised the mean almost $10,000 (about a 33% increase), which gives a distorted view of the data.

This is why it is also important to indicate how spread out the data are. The most basic measure of spread is the range. The *range* of a data set is the difference between the largest value and the smallest value. For the original 6 employees, the range was $13,000. When the seventh employee (the boss's grandson, perhaps?) was hired, the range jumped to $75,000.

When the data fit the bell curve, the measure of spread is called the standard deviation. In simplest terms, the *standard deviation* is the average difference between each data point and the mean of the data. For normal distributions, the rule is that 68% of the data values will lie between a point 1 standard deviation below the mean and a point 1 standard deviation above the mean, 95% of the data values will lie between a point 2 standard deviations below the mean and a point 2 standard deviations above the mean, and 99.5% of the data values will lie between a point 3 standard deviations below the mean and a point 3 standard deviations above the mean.

A manufacturer claims the life span of its 60-Watt bulbs is normally distributed with a mean of 760 hours and a standard deviation of 30 hours. The distribution of the life spans for light bulbs should look like this.

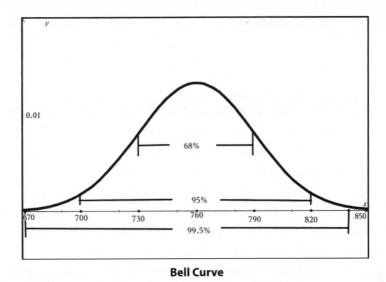

Bell Curve

68% of the light bulbs will have a life span between 730 and 790 hours, 95% of the light bulbs will last between 700 and 820 hours, and 99.5% of the light bulbs will last between 670 and 850 hours.

ESSENTIAL

Many companies use statistics to measure how well they are performing. General Electric introduced a concept they called 6 Sigma using the idea of the bell curve (the graphical representation of the normal distribution) to keep its employees focused on maintaining excellence in their performance.

Polls

News media often cite the results of polls conducted on issues that have local or national significance. One example is a May 2, 2013, release from the Quinnipiac University Poll service (*www.quinnipiac.edu/institutes-and -centers/polling-institute/national/release-detail?ReleaseID=1891*). Item number 27 on that web page reads as follows:

"Do you think items and services purchased on the Internet should be subject to state sales taxes or not?"

Here are the results recorded by political affiliation and by race:

	Total	Rep	Dem	Ind	Men	Wom	Wht	Blk	Hsp
Yes	37%	33%	46%	33%	38%	36%	37%	37%	45%
No	56	61	45	60	55	56	56	52	48
DK/NA	8	6	9	7	7	9	7	11	7

Here are the results recorded by college degree, household income, and age:

	COLLEGE DEG		ANNUAL HSHOLD INC			AGE (in years)			
	Yes	No	<50K	50–100	>100K	18–29	30–44	45–64	65+
Yes	38%	36%	38%	33%	40%	29%	30%	39%	49%
No	55	56	54	62	55	68	64	55	37
DK/NA	7	8	8	5	5	3	6	7	15

Political figures gather this kind of information to help them understand how their constituents would like them to vote. (Indeed, some cynics claim that many politicians will not take any political stance until they see poll results.)

How are polls conducted, and how valid is the information they convey? Polling services such as Quinnipiac and Gallup go to great lengths to ensure that their polls are as unbiased as they can possibly be. When you see the result of a poll on television, or in the newspaper, the phrase "margin of error $\pm 3\%$" is usually included to help readers understand what degree of confidence they can have in the validity of the results reported. It is financially and logistically unrealistic for every person to be polled. Statistically accurate strategies for whom to poll and where to poll them are applied to ensure as accurate an assessment as possible, but there is always the possibility that an error has occurred. The poll results given here suggest that 55% of people with a college degree are opposed to charging a sales tax on purchases of goods sold via the Internet. Is that number exactly right? No, probably not. Is it close? Yes. Probably.

Beware of results conducted by local organizations. Such organizations are not necessarily disreputable, but there have been way too many instances of "lies and damned lies" using statistics. For example, if the data collected regarding a local community issue is taken on a Wednesday from 11 A.M. to 3 P.M. because that is when the pollsters were available, then you can imagine the number of constituents they missed because those people were at work.

ESSENTIAL

In a campaign ad in the 1970s, the manufacturers of Trident gum announced that "4 out of 5 dentists recommend if you chew gum, you should chew sugarless gum." This was at a time when sugarless food products were just beginning to appear on the market. Today, a glance at most checkout counters in a supermarket will show that there are as least as many sugarless gum products available as those containing sugar.

Sadly, some misrepresentations of data are intentional. Ben Goldacre, a British physician, has reported in the series of TEDTalks of companies that intentionally misrepresent their studies and/or their data in the field of medical research in such a way that doctors are unaware of the errors.

This does not mean that you should doubt every conclusion drawn from a statistical study. It does mean that should you suspect that something is not quite right about what is being reported and you have the wherewithal to investigate how the findings were found, you can ask for how the data was collected or even possibly a copy of the data.

QUESTION

When do you use probability and statistics?
Consider when you measure your gas mileage on your last fill-up and compare it to the numbers ratings stated when you bought your car. You might also consider the average amount of time it takes you to commute to work. What else did you add to your list of when you use math?

Exercises

Answer the following questions about probability and statistics.

1. Two fair dice are rolled, and the sum is computed. What is the probability that the sum is greater than 7?
2. Two cards are drawn from a well-shuffled bridge deck, with the first card not being returned to the deck before the second card is selected. What is the probability that both cards are face cards (picture cards)?
3. The New York State Lotto has the player select 6 numbers from 1 through 54. What is the probability of selecting the winning numbers for the grand prize?
4. The salaries for 22 active players on the New York Yankees' roster at the beginning of the 2013 season are given. Calculate the median and the mean for these salaries.

$490,525	$935,000	$3,100,000	$15,000,000
$512,425	$1,200,000	$3,150,000	$15,000,000
$515,100	$1,500,000	$6,500,000	$24,285,714
$533,300	$1,875,000	$7,150,000	$24,642,857
$850,000	$2,000,000	$10,000,000	
$900,000	$2,850,000	$12,000,000	

Source: *espn.go.com/mlb/team/salaries/_/name/nyy/new-york-yankees*

5. Find the mean and the median for the number of hours of overtime (OT) paid in a year, as recorded in the following table.

Hours	Frequency
1	40
2	32
3	40
4	28
5	19
7	10

CHAPTER 16

Math in the Home

There are a multitude of math problems that arise in the home. You have already read about math in the kitchen, but this chapter examines the mathematics of maintaining your house or apartment. Most of the formulas come from middle-school geometry, so there is no need to worry that they will be too complicated.

Area and Volume Formulas You Should Know

There are three sets of formulas that you should be able to use (even if you have to refer to this text or go online to recall them): the Pythagorean theorem, area formulas for simple geometric figures, and volume formulas for prisms, cylinders, and cones.

The Pythagorean Theorem

In any right triangle (a triangle that includes a 90° angle), the longest side of the triangle is called the hypotenuse, while the two sides that make up the right (90°) angle are called the legs. Traditionally, the legs are labeled a and b, and the hypotenuse is labeled c.

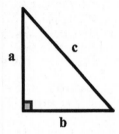

Hypotenuse and Legs of a Triangle.

The *Pythagorean theorem* states that in every right triangle, $a^2 + b^2 = c^2$ (that is, the sum of the squares of the lengths of the legs is equal to the square of the length of the hypotenuse).

FACT

Squaring a number means multiplying the number by itself. The square of 7, which is written as 7^2, is $7 \times 7 = 49$. The reverse of this process is called finding the square root. The square root of 49 is 7. The square root can be found on a calculator by pressing the square root key ($\sqrt{\ }$) if your calculator has one, or by using the exponent (power) function (represented by the caret ∧) with the number 0.5. That is, $49 \wedge 0.5 = 7$.

Example: Find the length of the hypotenuse of a right triangle if the legs have lengths of 5 in. and 12 in.

Use the theorem to write $5^2 + 12^2 = c^2$.
After squaring 5 and 12: $25 + 144 = c^2$
Add the two numbers: $169 = c^2$
The square root of 169 is 13, so the length of the hypotenuse is 13 in.

FACT

Records have shown that both the Egyptians and the Chinese knew of the Pythagorean theorem long before Pythagoras was born. Both of those civilizations were primarily interested in the practical applications of the relationship. The theorem is named after Pythagoras because he supplied a proof of the relationship—and because most of what we study in geometry in based on the work of the ancient Greeks.

Example: Find the length of a leg of a right triangle where the other leg has a length of 23 cm and the hypotenuse has a length of 35 cm.

Use the theorem to write $a^2 + 23^2 = 35^2$.
After squaring 23 and 35: $a^2 + 529 = 1225$
Subtract 529 from both sides: $a^2 = 696$
Use your calculator to get the square root: $a = \sqrt{696} = 696 \wedge 0.5$, which is approximately 26.4 cm.

Area Formulas

The basic *area formulas* from geometry that are worth knowing are those for the square, rectangle, triangle, parallelogram, trapezoid, and circle.

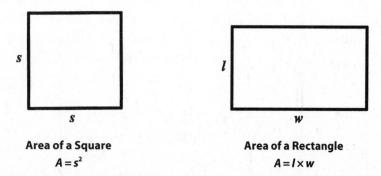

Area of a Square
$A = s^2$

Area of a Rectangle
$A = l \times w$

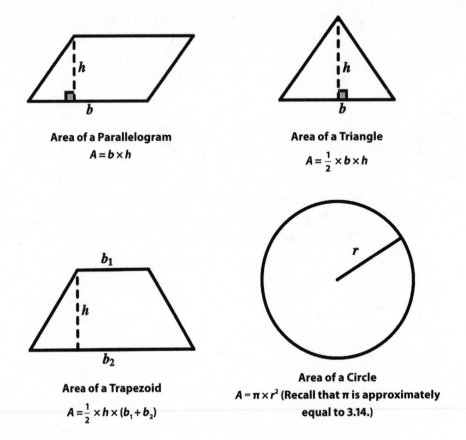

Area of a Parallelogram
$$A = b \times h$$

Area of a Triangle
$$A = \frac{1}{2} \times b \times h$$

Area of a Trapezoid
$$A = \frac{1}{2} \times h \times (b_1 + b_2)$$

Area of a Circle
$A = \pi \times r^2$ (Recall that π is approximately equal to 3.14.)

Solving Problems

Find the area for each of the figures in the following problems.

1. Find the area of a square that has sides with a length of 3 feet.

   The area of the square is found by squaring 3 ft., so the answer is 9 square feet. This can be written as 9 sq. ft. or as 9 ft². It is important to note that this same square also measures 1 yard on each side, so the area of the square is 1 sq. yd. Although there are 3 ft. in 1 yd., there are 9 square feet in 1 square yard.

2. The area of a rectangle is 120 sq. ft. Find the length of one side of the rectangle if the length of an adjacent side is 15 ft.

The area of the rectangle is $l \times w$, or length × width. Substitute 15 for l, the length, to write the equation $15 \times w = 120$. Divide by 15 to find that the length of the adjacent side, which we have designated the width, is 8 ft.

3. Find the area of the parallelogram shown.

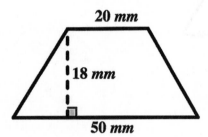

25 cm   20 cm   40 cm

The height is always drawn to the base, so the height is 20 cm and the base is 40 cm. The area of the parallelogram is 800 sq. cm.

4. Find the area of the trapezoid shown.

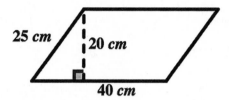

20 mm   18 mm   50 mm

The two bases have lengths of 20 mm and 50 mm, and the height of the trapezoid is 18 mm. The area of the trapezoid is $\frac{1}{2} \times 18 \times (20 + 50) = 9 \times 70 = 630$ sq. mm.

5. Find the area of a circle that has a radius of 4 ft.

The formula for the area of the circle is $\pi \times r^2$, so the area is $\pi \times 4^2 = 16\pi$ sq. ft. As nice an answer as that is, $16 \times 3.14 = 50.24$ ft^2 is a more practical way to express it.

ESSENTIAL

Area measurements are given in square units. When you multiply feet by feet, the answer is in square feet.

Volume

There are essentially three shapes that make up the basic solids whose volume formulas are known: the prism, the pyramid, and the sphere. The prism is a solid whose top and bottom (the bases) are exactly the same shape and size. The pyramid can have any shape as the base, but all the lateral edges come to a point (similar to the pyramids in Egypt—the difference between the pyramids in Egypt and the mathematical definition of a pyramid is that in math the base does not have to be a rectangle). The sphere is a ball (or a three-dimensional circle, if you prefer).

Three special cases of solids are usually identified. A prism with all sides of equal length, and all angles right angles, is called a *cube*. A prism with a circular base is called a *cylinder*. A pyramid with a circular base is called a *cone*. Because the base for the prism and the pyramid can be any shape, the formula for the volume is given in terms of the area of the base, B.

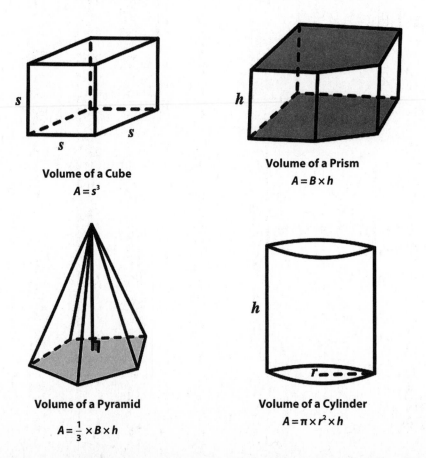

Volume of a Cube
$A = s^3$

Volume of a Prism
$A = B \times h$

Volume of a Pyramid
$A = \frac{1}{3} \times B \times h$

Volume of a Cylinder
$A = \pi \times r^2 \times h$

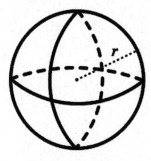

Volume of a Sphere

$$A = \frac{4}{3} \times \pi \times r^3$$

Volume of a Cone

$$A = \frac{1}{3} \times \pi \times r^2 \times h$$

Whereas area is measured in square units, volume is measured in cubic units.

Example: Find the volume of a cube that measures 5 cm on each side.

The volume of a cube is s^3, so the volume of this cube is $5 \times 5 \times 5 = 125$ cubic centimeters, or 125 cu. cm, or 125 cm^3. (In the medical field, the cubic centimeter is often abbreviated as cc.)

Example: The base of a rectangular box has dimensions 8 in. by 12 in. What is the height of the box if the volume of the box is 480 in^3?

The base of the box is a rectangle, and the area of a rectangle is $l \times w$. For this box, the area of the base is $8 \times 12 = 96$ in^2. Solve $96 \times h = 480$ to get $h = 5$ in.

Example: You have two pitchers that are both in the shape of a cylinder. One of the pitchers has a radius of 4 in. and a height of 8 in. The other pitcher has a radius of 3 in. and a height of 15 in. Which pitcher has the greater capacity?

The first pitcher has a volume of $\pi \times 4^2 \times 8 = 402.1$ in^3. and the other pitcher has a volume of $\pi \times 3^2 \times 15 = 339.3$ in^3. The shorter pitcher has the greater volume.

Example: While he was sleeping, a snowstorm left a few inches of snow in Colin's yard. It was the wet, heavy snow that is perfect for making a snowman. Colin and his father go out to the yard and roll the snow into a shape that is approximately spherical with a radius of 60 cm. The second snowball, also in a shape approximating a sphere, has a radius of 45 cm. Dad is a little surprised when he tries to lift the second snowball to put it on top of the first. He uses his smartphone to find the density of wet snow for his area and finds that it is roughly 140 kg/m³ (kilograms per cubic meter). Estimate the weight of the second snowball in both kilograms and pounds.

Because the density of the snow is given in cubic meters, convert the radius of the snowball from 45 cm to 0.45 m. The volume of the snowball (sphere) is approximately $\frac{4}{3} \times \pi \times (0.45)^3 = 0.381704$ m³. Multiply this volume by the density of the snow, and find that the snowball weighs $0.381704 \times 140 = 53.4$ kg. Since 1 kg is the equivalent of 2.2 pounds, the snowball weighs 117.6 lb. Colin and Dad might want to make a smaller snowman!

Painting the Walls

Maintaining the interior of your house is as important as maintaining the exterior. This means painting the walls on a regular basis. Although there is no exact schedule for when painting should be done, factors such as age of children, whether there is a fireplace, and exposure to direct sunlight affect the frequency with which a room should be painted. You will have to determine the amount of paint you should purchase, along with brushes, paint rollers, and floor covers (tarps), unless you use old sheets or blankets to protect your floor.

FACT

Computing the area of the walls that need to be painted or the area of the floors that need floor covering are some of the most common applications of basic geometry in the home.

When you go to the store to consider paint colors and paint prices, take a look at the paint can to find the area that a gallon of paint can cover. For most brands, that number should be approximately 400 ft² per gallon (or 37 m² per liter). Be warned that if the room you are painting has not been painted in a while, or if you are planning to paint a dark or bold-colored wall with a lighter color (or vice versa), you may need to use two coats of paint.

Example 1

Chris and Diane are going to paint their living room. The room measures 20' by 12' (20 ft. by 12 ft.) and has an 8-ft. ceiling. One of the shorter walls has no windows or doors in it, so the area of this wall is $12 \times 8 = 96$ sq. ft. One of the adjacent walls is part of the outside of the house and has two windows. The wall is 20 ft. long, and each of the windows (including the wood frame around the windows that will not be painted) measures 58.5" by 37" (58.5 inches by 37 inches).

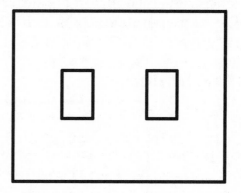

The painting area for this wall will be $20 \times 8 = 160$ sq. ft. minus the area of the two windows. The dimensions of the windows are awkward in the sense that these numbers do not convert easily to feet. You could, of course, find the exact areas of the windows by dividing each of the window dimensions by 12 and multiplying to get the area of each window in square feet, but all you really need is a rough estimate. The width of the window and frame, 37 in., can be rounded to 36 in., or 3 ft. The height of the window, 58.5 in., is almost 60 in., or 5 ft. The area of each window is approximately 15 sq. ft. Subtract this number twice (once for each window) from 160 to get the painting area for the wall, 30 sq. ft. (If you did the arithmetic for the exact window dimensions, you would find that your estimate was only 0.06 sq. ft. larger than the "true" answer.)

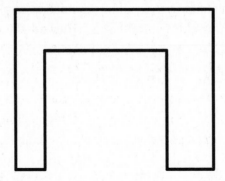

The second, longer wall has an entrance to the dining room. The doorway between the two rooms has dimensions 94" by 81". There are two doorjambs 6 in. wide that also need to be painted. Estimate the width of the doorway as 96 in., or 8 ft. The height of the door is 81 in., and that converts nicely to 6.75 ft. (81 divided by 12). The area of the wall formed by the door that will not be painted is $8 \times 6.75 = 54$ sq. ft. The area of the two doorjambs is $2 \times 6.75 \times 0.5 = 6.75$ sq. ft. The painting area for this wall (and doorjambs) is therefore $160 + 6.75 - 54 = 112.75$ sq. ft.

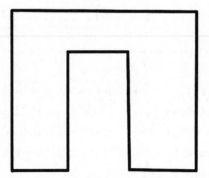

The fourth wall of this room has a doorway leading to the foyer. The wall measures 12' by 8', the doorway measures 4' by 81', and each doorjamb measures 6" by 81". The entire wall has an area of 96 sq. ft., the two doorjambs have an area of 6.75 sq. ft., and the doorway has an area of $4 \times 6.75 = 27$ sq. ft. The painting area for this wall is therefore $96 + 6.75 - 27 = 75.75$ sq. ft.

The total painting area for this room (not including the ceiling) is $96 + 130 + 112.75 + 75.75 = 414.5$ sq. ft. If Chris and Diane know that the walls have been painted regularly, they can assume they will need a gallon and a quart (the next smaller size in which their paint can be purchased). If the walls have not been maintained, they may need to buy 2 gallons of paint.

Example 2

Roger and Sue want to have their front room painted. This room is called a great room because of its high ceiling. One of the outside walls of the room has a panel of windows.

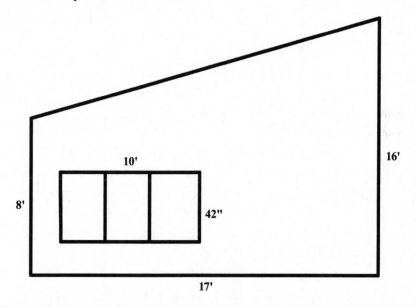

The painting area for this wall is the area of the trapezoid minus the area of the windows. The bases (the parallel sides) of the trapezoid measure 8 ft. and 16 ft., and the "height" of the trapezoid is the segment perpendicular to this floor; it measures 17 ft. The area of the trapezoid is $\frac{1}{2} \times 17 \times (8 + 16) = 204$ sq. ft. The height of the windows is 42 in., and this is equivalent to 3.5 ft. The area of the windows, then, is 35 sq. ft. The painting area of this wall is $204 - 35 = 169$ sq. ft.

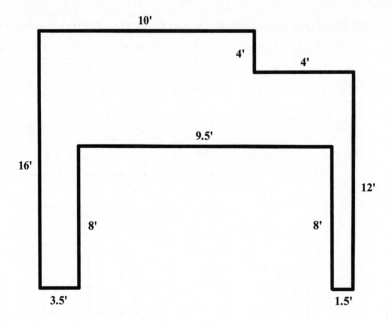

The wall joining the great room to the dining/recreational area is mostly open on the first floor. Part of this wall forms a wall for a bedroom on the second floor, and the remainder of the wall is a half-wall allowing a view down to the first floor. The painting area is a collection of rectangular regions that are placed together.

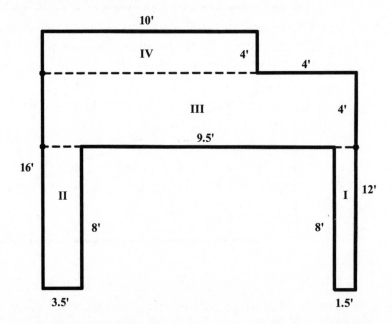

Region I is a rectangle with dimensions 1.5' × 8' for an area of 12 sq. ft.

Region II is a rectangle with dimensions 3.5' × 8' for an area of 28 sq. ft.

Region III is a rectangle with dimensions 4' × 9.5' for an area of 56 sq. ft.

Finally, region IV is a rectangle with dimensions 10' × 4' for an area of 40 sq. ft.

The total area for this wall is 12 + 28 + 56 + 40 = 136 sq. ft.

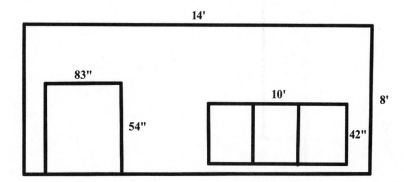

The wall at the front of the house also has a panel of windows, as well as the front door. The area of the region to be painted on this wall is computed by subtracting the area of the windows (35 sq. ft.) and the door frame (consisting of the door, a side window, and the frame). The door has a height of 83 in., which is almost 7 ft. (84 in.), and the width of the door is 54 in., or 4.5 ft. The area of the door is approximately 31.5 sq. ft. The painting area for this wall is 112 − 35 − 31.5 = 45.5 sq. ft.

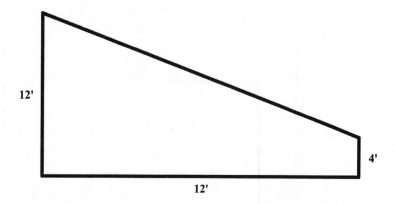

A half-wall in the shape of a trapezoid (because its shape must follow the staircase down to the floor below) and the wall bound the stairway on one side, and another wall from floor to ceiling will form the fourth wall of the great room. For the sake of simplicity, let's include the staircase in the area. (At worst, Roger and Sue will have a little paint left over because the area calculated is larger than the area they are painting.) The area of this trapezoid is $\frac{1}{2} \times 12 \times (4 + 12) = 96$ sq. ft.

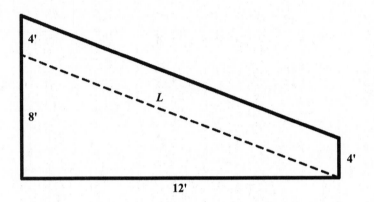

The inside of this wall must also be painted, along with the 6-ft. handrail and the frame at the bottom of the stairway. Use the right triangle and the Pythagorean theorem to find the length, L, of the handrail.

$$8^2 + 12^2 = L^2$$
$$64 + 144 = L^2$$
$$L^2 = 208$$

The length of the handrail is approximately 14.42 ft. (which you can round to 14.5 ft. for easy calculation). The area of the parallelogram with sides 14.5' by 4' is less than the area of the rectangle with sides 14.5' by 4'. Draw the parallelogram and drop the height from the vertex (point where the two sides meet) to the 14.5-ft. side. The side with a length of 4 ft. serves as the hypotenuse of the little right triangle formed, so the height of the parallelogram is less than 4 ft. The area of the inside of the wall can be estimated as $14.5 \times 4 = 58$ sq. ft. The handrail has an area $14.5 \times 0.5 = 7.25$ sq. ft., and the frame at the bottom of the stairway has an area of $4.5 \times 0.5 = 2.25$ sq. ft. The total area for the inside of this wall, the handrail, and the frame is 67.5 sq. ft.

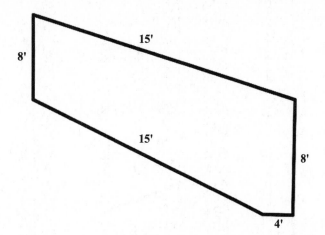

The area of the wall that forms the far side of the stairs can also be cut up into regions.

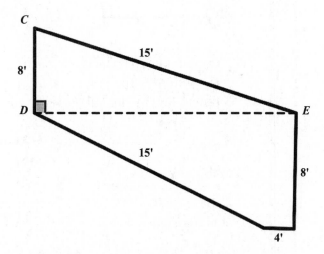

Triangle *CDE* is a right triangle. The length of side *DE* can be calculated using the Pythagorean theorem.

$$8^2 + DE^2 = 15^2$$
$$64 + DE^2 = 225$$
$$DE^2 = 161$$

Side *DE* is approximately 12.68 ft. long; round this to 13 ft. for an easier number to use. The area of triangle *CDE* is $\frac{1}{2} \times 13 \times 8 = 52$ sq. ft. The area of the trapezoid forming the rest of the wall is $\frac{1}{2} \times 8 \times (13 + 4) = 68$ sq. ft. The painting area for the far wall of the stairway is thus $52 + 68 = 120$ sq. ft.

Adding the areas of all these regions together gives a total area of $169 + 136 + 45.5 + 96 + 67.5 + 120 = 634$ sq. ft. Roger and Sue will need to purchase 2 gallons of paint to put a single coat of paint on the walls of the great room.

Example 3

Brendon and Stacey also have a great room. The wall forming the outside of the house has a fireplace, a Norman window, and two traditional rectangular windows.

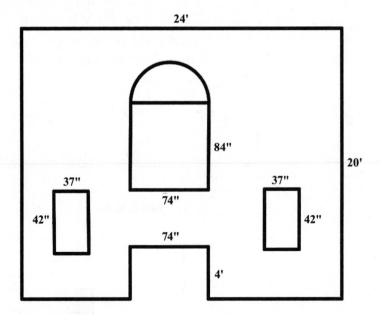

The painting area for this wall is 480 sq. ft., minus the area of the fireplace and the windows. The fireplace is a little over 6 ft. wide. Approximate the area of the fireplace as 24 sq. ft. The dimensions for each of the rectangular windows can be estimated as 3' by 4.5'. The area for both rectangular windows is $2 \times 3 \times 4.5 = 27$ sq. ft. The Norman window consists of a rectangle and a semicircle. The dimensions of the rectangular portion of the window are approximately 6' by 7', for a total area of 42 sq. ft. The radius of the semicircle portion of the window is 37 in., or approximately 3 ft. The area of the semicircle is half the area of a full circle with the same dimensions, so the area of this portion of the window is $\frac{1}{2} \times \pi \times 3^2 = 14.1$ sq. ft. The painting area for this wall is thus $480 - 27 - 42 - 14 = 397$ sq. ft.

Flooring

Whereas the amount of paint you need is measured in square feet (the information given on the can of paint indicates the coverage in square feet per gallon), flooring can be measured in a number of different ways. Carpeting is measured in square yards. Laminate flooring and bamboo flooring are measured in square feet. Tile flooring is measured on the basis of the size of the tile you use (9", 12", or 18"). When measuring, overestimate the area of the floor, rather than underestimating it. You can always cut off extra material, but discovering late in the job that you do not have enough carpeting or wood can be very costly.

Example

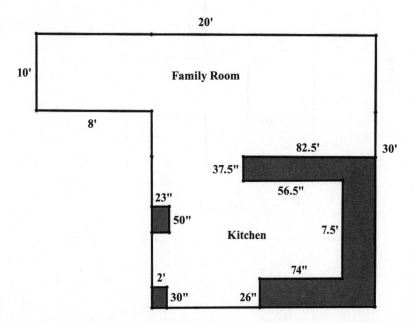

Diane and Chris plan to put new flooring in their combined family room and kitchen. As shown in the diagram, the family room has an L shape.

There are a built-in desk and a built-in pantry in the kitchen, as well as a counter in the shape of a U. The total area of the rooms can be calculated by breaking up the L from the family room and kitchen, then subtracting the areas occupied by the objects that cannot be moved. One plan is to use a laminate flooring for the two rooms.

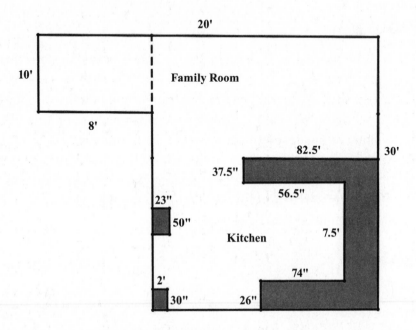

The small rectangle in the family room has an area of 80 sq. ft. The large rectangle that makes up the rest of the two rooms measures 12 ft. (20' − 8') by 30 ft., for an area of 360 sq. ft. Therefore, the area for the two rooms is 440 sq. ft. In order to overestimate the floor area, you will underestimate the area of the objects that cannot be moved (and thus will not have flooring installed under them). The desk with dimensions 23" by 50" has an area of about 1.5' × 4' = 6 sq. ft. (compared to 23 × 50 ÷ 144 = 7.98 sq. ft.). The pantry has an area of 2' × 2.5' = 5 sq. ft. The counter will be divided into three pieces. The first piece has dimensions 37.5" by 82.5". The area for this piece is estimated as 3' × 6' = 18 sq. ft. The second piece of the U measures 7.5 ft. long by 82.5" − 56.5" = 26" wide. The area estimate is 7.5' × 2' = 15 sq. ft. The last part of the U is 26 in. wide and 74" + 26" = 100" long. The estimate for the area of this section is 2' × 8' = 16 sq. ft. The total floor area that is not being included is 6 + 5 + 18 + 15 + 16 = 60 sq. ft., so the total area of the two rooms that will need new flooring is 440 − 60 = 380 sq. ft.

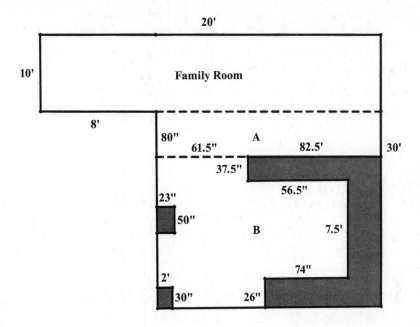

The second plan Diane and Chris are considering is to put a laminate flooring in the family room up to the wall forming the L, and to use tile for the kitchen area. If they choose to do this, they will need $10' \times 20' = 200$ sq. ft. of laminate flooring and $440 - 200 = 240$ sq. ft. of tile.

Diane and Chris need to consider that tiles come in various sizes. They do some shopping and find that square tiles are commonly available in five different sizes: 4-in., 6-in., 9-in., 12-in., and 18-in. tiles. To determine how many tiles they need to buy, they will have to divide the area of the floor by the area of each tile (measured in square feet). The 4-in. tile measures $\frac{1}{3}$ ft. per side and has an area of $\frac{1}{9}$ sq. ft. The 6-in. tile measures $\frac{1}{2}$ ft. per side and has an area of $\frac{1}{4}$ sq. ft. The 9-in. tile measures $\frac{3}{4}$ ft. per side and has an area of $\frac{9}{16}$ sq. ft. The 12-in. tile measures 1 ft. per side and has an area of 1 sq. ft. The 18-in. tile measures $\frac{3}{2}$ ft. per side and has an area of $\frac{9}{4}$ sq. ft. The table shows the number of tiles that Diane and Chris will need to purchase once they decide which size to buy.

Tile Size	Number Needed (240 ÷ area of 1 tile)
4-in.	3,840
6-in.	960
9-in.	427 (rounded up)
12-in.	240
18-in.	107 (rounded up)

FACT

A line of grout that is usually $\frac{1}{4}$ in. wide generally separates the tiles.

The living room in Chris and Diane's house measures 12' by 20'. They find a wall-to-wall rug that costs $21.42 per square yard. The area of the room (and of the rug) is 240 sq. ft., or 240 ÷ 9 = 26.667 sq. yd. The cost of the rug is 21.42 × 26.667 = $571.20.

New Curtains

Kate and Russ are hanging new curtains. The curtains are 84 in. long, and the curtain they decide on measures 50 in. long. The window over which these curtains are to hang is 37 in. wide and 58.5 in. tall. Because the curtain rod is fairly heavy, it will have to be supported with a bracket in the center of the window. The remaining brackets that will hold the curtain rod are to be placed 2.5 in. from each end of the curtain rod. The bottom of the window frame is 24 in. above the floor. The top of the window frame is 24 + 58.5 = 82.5 in. above the floor.

Russ pencils a light line on the wall 2.5 in. above the top of the window frame. He next measures a position 18.5 in. from each end of the window frame and marks where the center bracket will go. He then measures 22.5 in. to the left and right of the center point and marks the wall where the end brackets will be located to hold the curtain rod. He grasps the curtain in the center and holds it in place so that he and Kate can take one last look before he drills holes in the wall. (The first rule in construction, for professionals and do-it-yourselfers alike, is "Measure twice, cut once.")

Kilowatt-Hours

A watt is a measure of power used to operate an electronic device. One thousand watts (1,000 W) is equal to 1 kilowatt (1 kW). The energy needed to produce this power is what you pay for each month when your electric bill comes. You are billed for electrical energy in terms of how many kilowatt-hours (kWh) you use. To explore this relationship, let's consider the light bulbs you use at home and how long you use them.

Calculating Cost

A 100-W (0.1-kW) bulb, when used for 1 hour, consumes $0.1 \text{ kW} \times 1 \text{ h} = 0.1$ kWh. The lifespan for light bulbs is approximately 740 hours, so the amount of energy needed to power the bulb over its lifespan is $0.1 \text{ kW} \times 740 \text{ h} = 74$ kWh. A number of companies now make compact fluorescent light (CFL) bulbs that are more energy-efficient while producing the same amount of light (measured in lumens) as conventional light bulbs. For example, Sylvania makes a 60-W replacement CFL bulb that uses 13 W and has a lifespan of 12,000 hours. Consider the energy savings—a difference of 47 W (0.047 kW) for a period of 12,000 hours (that is 500 full days) comes to $0.047 \text{kW} \times 12,000 \text{ h} = 564$ kWh. At roughly 12 cents per kilowatt-hour (the national average in October 2011, according to the National Public Radio show *Planet Money, www.npr.org*), using a CFL bulb yields savings of $67.68 for the life of the bulb—and that is just one light bulb! (It is true that CFL bulbs are more expensive to purchase, but you would need to buy $12,000 \div 740 = 17$ conventional light bulbs during the same time period in which you would use up a single CFL bulb .)

All your electrical devices list the specifications for the number of watts that they use. To calculate the amount of energy each uses, multiply the watts by the number of hours the device is used and divide by 1,000 (to convert watt-hours to kilowatt-hours). Be aware that some appliances that are always "on," such as the refrigerator, really are not always on. The compressor for the refrigerator runs when there is a need to cool the refrigerator. If you are listening, you can hear even the quietest refrigerator when the compressor turns on. If you are interested, keep a chart of the amount of time that the compressor runs during the day to gauge how much energy the refrigerator is using. The United States Department of

Energy website *www.energy.gov* offers a great deal of information about energy costs and tips for saving energy (and money).

The Water Bill

Many communities include a charge for sewage removal and processing on the water bill—think of it as the fee to bring it in and the fee to take it out. There is a water meter either inside or outside your house to record the amount of water used each time period; (for some that would be monthly, for others it would be bimonthly (every 2 months), and some pay quarterly). Knowing how much you pay each month certainly helps with budgeting. And more important, an unusually large change that cannot be accounted for by a big party or an extended visit from houseguests can be an indication that there is a leak somewhere—perhaps a leaking faucet, an improper seal on a toilet, or something more serious. If you notice a large increase in water usage, you should check for leaks before any water damage gets too serious. If you have a pool, you will want to let your water authority know when you are filling the pool or to arrange for a separate meter so that you will not be charged a sewage fee for the water used in the pool.

Taking Care of the Lawn and Garden

Each spring, homeowners across the country get their lawns ready for the new season. Some homeowners do the job themselves, and others hire a company to do the work. In either case, knowing the area of your lawn is important so that you can purchase the correct amount of fertilizer.

Calculating the Area of Your Lawn

Luis and Juanita are estimating the area of their lawn. The property map shows that even though the perimeter of the property (the distance around it) includes a curved corner, the shape of the property is roughly a trapezoid with bases 118 ft. and 70 ft. and a height of 180 ft. (estimated by extending the straight boundaries through the curve).

Many people save on their water bill by installing a well point to use to water their lawn rather instead of paying the expense of municipal water. Others incorporate a second water meter to avoid having to pay for the sewage charge on the water bill.

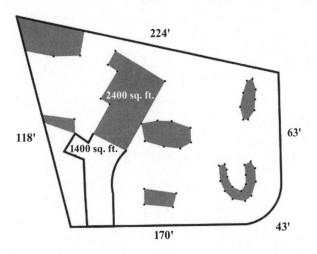

The area of the entire property is $\frac{1}{2} \times 180 \times (118 + 70) = 16{,}920$ sq. ft. The house has an area of 2,400 sq. ft., the driveway has an area of 1,400 sq. ft., and the six gardens that Juanita likes to work in have a combined area of 2,900 sq. ft. The rest of the area is thus $16{,}920 - (2400 + 1400 + 2900) = 10{,}220$ sq. ft. Luis and Juanita will need to buy enough fertilizer to cover this area (again, they will consult the label on the product they choose to learn how many square feet one package will cover).

Gardeners will buy mulch to help fertilize their gardens. Mulch delivered from a gardening store is measured in cubic yards. If the mulch is placed in the garden at a depth of 3 inches (or $\frac{1}{12}$ of a yard), one cubic yard of mulch can cover 12 square yards, or 108 square feet of garden.

How Tall Is That Tree?

A very practical application of proportions (and of similar triangles from your days in geometry class) is to find the height of a tall object, such as a tree, that it is not practical to measure directly. This method relies on measuring a shadow cast by this tall object. It is most accurate if the shadow falls on a uniformly horizontal surface (such as a walkway or lawn). But that rarely happens in the real world, and the approach you are about to learn will give you a good approximation if the slope of the surface is not too steep.

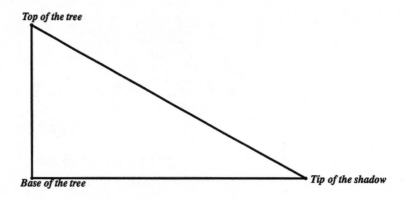

How to Approximate Height

Say you want to find the height of a tree near your home, just to be sure that it wouldn't hit your bedroom if a hurricane blew it down directly toward the house. You use a tape measure to measure the length of the shadow of the tree, and find that it is 52 ft. long from the base of the tree to the tip of the shadow.

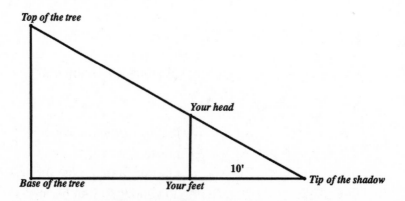

With no delay, step into the shadow of the tree, and position yourself so that the tip of your own shadow is in the same spot as the tip of the shadow made by the tree. Mark the point where you are standing, and measure the distance from there to the tip of the tree shadow. Say this distance is 10 ft. The last piece of information you need is your height. For the purpose of showing you how to finish the problem, it will be assumed that you are 6 ft. tall. (If you are really 5'8" or 6'3", you will still get a good approximation. If you are much shorter or taller than that, you may want to round your height to the nearest foot below or above 6 ft.) Use the proportion:

$$\frac{\text{Length of tree's shadow}}{\text{Height of tree}} = \frac{\text{Length of your shadow}}{\text{Your height}}$$

Using the information for this problem yields the proportion

$$\frac{52}{h} = \frac{10}{6}$$

Cross-multiply to get $10 \times h = 52 \times 6 = 312$. Divide by 10 to find that the tree is 31.2 ft. tall. That's about how far your house can be from this tree and not get even a scratch if it blows down—this year, anyway.

QUESTION

When do you use math around the house?
Consider the amount of money you spend on fertilizer for your lawn and garden or savings per year on electricity when you use energy-efficient appliances. What else did you add to your list of when you use math?

Exercises

Answer the following questions that involve maintaining and decorating a home.

1. The two shorter sides of a right triangle have lengths 7 cm and 24 cm. Find the length of the hypotenuse.
2. A craft project instructs you to take a piece of card stock with dimensions 12" by 20", roll it into a cylinder so that the circumference of the

circular base is the 20-in. side, and tape the edges together. What is the radius of the cylinder? What is the volume of the cylinder?

3. Find the volume of a rectangular box that measures 9" × 17" × 10".

4. Don and Cheryl have a wall in their great room that needs to be painted. Find the area of the wall that needs painting.

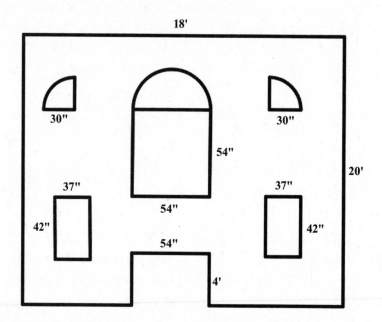

5. Patty and Dan are hanging curtains to cover their windows, which measure 47" wide by 62" tall. The bottom of the window frame is 18 in. from the floor. They buy curtains that are 82 in. long and a curtain rod that is 52 in. long. The rod does not need a center bracket, but Patty and Dan do want the rod to be centered on the window, and they want each supporting bracket to be $2\frac{1}{4}$ in. from its end of the curtain rod. Where should they drill the holes for the curtain rod?

6. Each morning, Diane and Chris use their Sunbeam coffeemaker to make 10 cups of coffee. The coffeemaker is rated at 900 W, and the rule of thumb has been that the coil works at maximum power for a number of minutes equal to the number of cups of coffee being made. If they are home approximately 320 days each year (when they are not visiting family members or on a vacation), how much do they pay to brew their

coffee in a year? (The price of 1 kWh of electrical energy in their area is 18.1 cents.)

7. John and Suzanne have a rectangular lot with a house and driveway on it. The rest of the property is lawn. Find the area of the lawn.

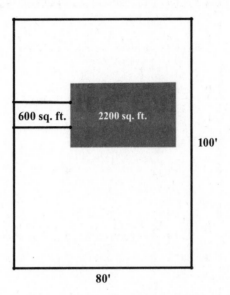

600 sq. ft.   2200 sq. ft.

100'

80'

8. Andrew, who is 6 ft. 2 in. tall, is using the shadow technique to find the height of a tree in his backyard. During the early afternoon, the tree casts a shadow 10 ft. long. Andrew steps into the shadow and positions himself so that the tip of his shadow coincides with the tip of the tree's shadow. Andrew then measures the distance from his position to the tip of the shadow and finds that it is 2.5 ft. Estimate the height of the tree.

CHAPTER 17

Math on the Job

The job that does not involve using math is rare indeed. It is not surprising that accountants, engineers, medical professionals, and banking personnel use math every day; people always associate these professions with needing good "number sense." Salespeople who compute percent tax or percent sales, administrative assistants who determine the amount of time it will take them to type a document from the number of words they can type per minute, and lawn service representatives quoting a price based on the area of the lawn—all use math on a regular basis. Many people "just do their job" and do not recognize that decisions they make are based on mathematics principles that have become ingrained in their everyday work.

You may not be in the same business as the people described in this chapter. However, as you read through the chapter, see if you recognize similarities to what you do. In addition, you may be asked someday who uses math in the "real" world. Math teachers hear this question much too often, and it is probably fair to say that you might have said those words when you were in school. Should you hear the question in the future, you now have some answers.

Window Replacement

Dick Vaillancourt, owner of Saratoga Springs Windows, says that aside from the obvious application of measuring windows, he needs to be constantly aware of the cost of materials and labor for installing windows, of the amount of money he spends on gasoline traveling to prospective clients (and consequently how well his vehicle performs), and of achieving a profit margin that enables him to stay in business. While some of the costs are based on flat-rate fees, others are based on percentages, and he needs to be able to compute these numbers. Like most people, Dick uses a calculator to do this.

Small-business owners know that aside from charging the cost of materials and labor to complete a project, they must charge enough money so that they can be paid and also have the capital available to hold them over when business is slow or to pay for upgrades in equipment. To accomplish this, they might add a percentage, sometimes called a markup, to the cost of the materials. For example, if the business owner pays $210 for materials and works with a 25% markup, he will charge his customer $210 + 25% of $210 (or $52.50) = $262.50 for the materials.

ESSENTIAL

People who work in sales, who are engineers, or who are bankers are not the only people who use math in their jobs. Many occupations use mathematics in ways that are so common that people often take them for granted.

Physical Therapy

Tracy Sherman of LaMarco Physical Therapy in Saratoga Springs, New York, and Kellie Hummel of Franziska Racker Centers and Tompkins County Department of Health Early Intervention Program in Ithaca, New York, are both physical therapists.

FACT

A physical therapist works with people who are recuperating from an injury or surgery to repair a bone or muscle. Among other things, a physical therapist will help patients regain the strength in their muscles or help increase the range of motion in their joints (e.g., how far you can bend your knee or raise your arm above your head).

They both say that most of the math they do was done while they were studying physics for their degree. They do not do the calculations while on the job, but refer back to physics to help them do their job.

While at work, they use specific tools to measure the angles formed by joints. (Taking such measurements is called goniometry, and the tool looks very much like a large protractor from geometry class.) These measurements are crucial during rehabilitative therapy for problems with the knees, for example, and when measuring patients for adaptive equipment.

When Tracy was asked how she determined what amount of weight to have her patients use when exercising, she answered, "It really is a guesstimate based on the patient. You do not ask someone in their late sixties to use as much weight on a leg press as you might ask of a twenty-year-old. In either case, you start with a reasonable weight and adjust as the patient needs it." She also said, "The initial weight I ask a patient to start with depends on the nature of the injury. There is a difference between treating a person experiencing an injury for the first time and someone who has been dealing with a recurring issue. Decisions are also based on whether the muscles you are working on impact one joint or two joints."

Exercise Physiologists

Nick, Mike, and Erin are exercise physiologists at Saratoga Health and Wellness, and they work with their customers in a bright and airy, open floor space. When asked how they use math when interacting with their customers, they came up with a number of interesting examples. For instance, one of the key components in exercising is reaching and maintaining the appropriate heart rate for your age and ability. An excerpt from Nick's blog on the company's website (*www.saratogahealthandwellness.com*) on April 30, 2013, reads as follows:

"One of the simplest methods of setting target heart rate range uses a straight percentage of the maximal heart rate. We, as clinicians, use 70% to 85% of an individual's heart rate max as the prescribed exercise intensity. This range of exercise intensities approximates 55% to 75% of your VO_2 (the volume of oxygen used during exercise) max and provides the stimulus needed to improve your cardiovascular potential. The oxygen in your bloodstream is used by your body to convert the energy from the food you eat into the energy molecules, called adenosine triphosphate (ATP), which your body uses at the cellular level.

"It is also simple to compute," writes Nick. "To calculate your target heart rate we first must calculate your estimated maximal heart rate. For this, we simply subtract your age from 220. Once we have your maximal heart rate, we simply multiply by the 70% and 85%."

For example, Nick explains, "I am 35 years old. My estimated maximal heart rate is 185 beats per minute (220 − 35 = 185). With a maximal heart rate of 185 beats per minute, my target heart rate would calculate out to 130 to 157 beats/min (185[0.70] = 130, 185[0.85] = 157)."

ESSENTIAL

One Metabolic Equivalent (1 MET) is 3.5 mL oxygen per kg body weight per minute. Put simply, METs measure how hard your body is working. At complete rest, your body uses 1 MET. A rule of thumb for computing your optimum MET level during exercise is to take 0.1 times your age and subtract it from 14.7.

When they were asked, "Why is it common to lift weights with 10 repetitions per set? Why not a different number of repetitions?" each staff member gave the same response: "The idea, when lifting weights, is to fatigue the muscles. Ten is not a magic number, but just a common application. One person might do three sets of 10 while another might do 2 sets of 15 repetitions. They are both doing 30 reps but one needs more reps to tire the muscles."

Business Calculations

Some of the math that Nick, Mike, and Erin do isn't directly tied to exercise. As owners of the business, they must attend to a wide variety of money matters. The insurance they carry costs 30 times as much as the rent they pay for the facility! In determining the fee for membership, they must take into account these costs (as well as their own salaries, the salaries of their employees, and the costs of maintaining their equipment). Part of what they do is to estimate the number of customers per square foot for their facility and the rate at which they hope the facility will be used. Mike indicated that he was in the process of making a proposal to a local business that has approximately 200 employees. "Based on past experience, we know that maybe 15% of the employees would take advantage of a membership to this facility. If we offered free use to any employee of the company for an annual flat rate paid by the company, we would have a win-win situation. The company can offer a way for its employees to maintain their health and we would have an increase in income." Using past experience—that is, analyzing data from a previous situation and applying them to the current situation—is part of the decision-making process involving statistics.

As with the ladies who discussed their work in Physical Therapy, past practice still works but can be adapted as needed for an individual as opposed to a hard-and-fast rule.

Taking Care of the Pool

Gary Torrisi, owner of Apple Pools in Saratoga Springs, New York, uses mathematics in a variety of ways in his business life. For each installation contract, he first uses geometry as he works with the homeowner to lay out the pool. Since many municipalities have different rules for how far a pool must be from property lines, he measures the required distances so that he and the homeowner can determine where the pool can be placed. The outlines for rectangular, circular, and oval pools can be easily marked off in the available space. Some pools are shaped like this:

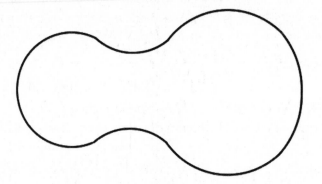

These pools require a great deal of measuring to ensure an accurate fit. Two points, marked A and B on the diagram, are placed 15 feet apart. A series of pegs are placed in the ground to form the outline of the pool, and the distances from points A and B are computed. These measurements can then be used to determine the perimeter of the pool and to find the measurements of the pool lining that will be used.

A lot of the mathematics that is associated with owning a pool comes from middle school. Determining distances from lot lines and homes to meet community ordinances, measuring diameters for piping, and calculating ratios of chemicals to insure proper water balance are chief among these applications.

The volume of the water in the pool can be computed using the formulas for surface area and volume. Circular pools usually have a constant depth, so the volume of the water in the pool can be calculated as $\pi r^2 h$, where r is the radius of the pool and h is the depth of the water. Rectangular pools usually have a gradual slope from the shallow end to the deep end. Use the average of the depth at the shallow end and the depth at the deep end as the depth of the pool, and multiply this number by the surface area of the water. For the irregularly shaped pool shown, the surface area can be computed by the formula, Surface area $= (A + B) \times L \times 0.45$.

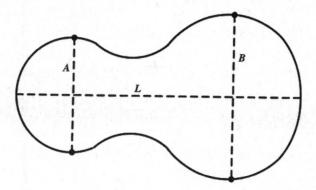

The volume of the water is computed using the constant depth of the water in a round pool or, for a pool with a gradual slope, the average of the depth at the shallow end and the depth at the deep end. These volumes are all measured in cubic feet. Because 1 cubic foot = about 7.48 gallons, multiply the number of cubic feet by 7.5 to estimate the number of gallons of water that will be needed to fill the pool. The homeowner will need to know the cost of filling the pool. If the pool is being filled with water from the municipality, the owner should check with the municipality, because most bills for water usage also include sewer usage. The owner will not want to pay this added fee. The alternative is to hire a private company to deliver water to the site to fill the pool.

Once the pool has been installed and filled with water, the issue of maintaining the pool must be considered. Water filtration involves the use of appropriate chemicals and a water pump. "Getting an efficient pump can save on electricity costs, and working with a pool professional to ensure that the water chemistry is appropriate for your area is not as expensive as one might think," offers Mr. Torrisi.

Shoe Manufacturing

The staff at Vasyli, an international footcare and footwear retailer and wholesaler out of San Rafael, California, use math differently from each other. Some of the employees use geometry (especially the designers), while others need to apply their algebra skills. Almost all of the decisions made are done so with the statistical data available.

Tony, a designer for the company, wrote, "Well, I do a fair amount of work in 2D specs and 3D CAD. I take measurements and make conversions between inches, decimals, and millimeters. I use some geometry when building images, and thinking of ways for symmetry, and repetition. In 3D I'm using basic math and some differential geometry for evaluation, etc."

Mark, another designer with the company, wrote, "One instance … I was designing a hexagon pattern in Illustrator that required equal spacing. So I did basic algebra to figure out the spacing within the parameters of the hex."

Steve is an executive vice president and general manager. He wrote. "I use math for calculating prices to customers after a discount.

- Margins on new shoes;
- Sales forecasts by customer and product;
- Stock turns and sell through of our retailers.

"We do, indeed, use math a lot. When sending new shoe designs to factories, math is our common language. We show measurements in millimeters for each part of the shoe's components, including outsole, midsole, upper, and trim designs. Trims are shown in a 1:1 scale so the China team can accurately create a model for the new development.

"Math is also used when calculating costs for a shoe design. This especially comes into play when we have to reach tight margins on a new shoe. The Sales Department gives us a MSRP (manufacturer's suggested retail price) we need to hit, and it's up to us to design a shoe that can be made for the right price. This means calculating any new tooling developments, labor costs, and materials per square foot.

"This may be more geometry, but a big part of costing comes from leather usage. The upper patterns are cut from a leather hide. In 'offline' factories this means a worker manually places cutting dies on the hide as closely as possible before making the cut. In 'online' factories a machine does the calculations automatically. The more uppers that can be cut from a single hide the lower the costs will be for the materials."

Debra is the office manager. She wrote, "Lately I have had to use my math skills for the following: I have 3 rooms that require shelving and currently have: 11 shelves 60⅜" wide × 18.5" deep × 72.25" high and 2 shelves 48.75" wide × 18" deep × 75" wide. Using the square footage of each room, I have had to determine the best way to set up the existing shelves for maximum use."

Bruce is the chief operating officer of Vasyli. He wrote, "Our business runs on math. Even product, which is our currency, runs math. We use dimensions for design, numerical standard for Quality Control and product testing, math for budgeting forecasting and purchasing. We even use math for customer ratings. Our success or failure is measured in numbers."

Shaun, the product development director, provided the following list of how he uses math.

Professionally:

- Costing and profit margins
- Product and production flow
- Replenishment lead times
- Retail math
- Budget
- Travel costs
- Cost breakdowns—material consumption, LOP, material costing
- Allocation and factory planning; capacity performance/analysis/consolidation & expansion needs

- Vendor performance
- Rating reviews; human resource performance reviews; salaries; bonuses

Kristen Monahan is the supply chain manager. She uses an Excel spreadsheet to analyze data as she forecasts the styles, colors, and sizes of shoes that will be ordered for the next season. The accuracy of these forecasts plays a very large part in the failure or success of the company. In this example, forecasting is part science and part art. Determining colors and styles for the upcoming season is more artistic in nature but knowing who will buy them comes from a statistical analysis and working closely with sales reps in the field.

QUESTION

When do you use math at work?
Consider the amount of money you spend commuting to work each day.
What else did you add to your list of when you use math?

CHAPTER 18

Playing Games

Mathematical principles underlie almost every game you play. At the very least, you use math when you develop your playing strategy, which you base on the proficiency of your opponent as well as on what has transpired up to some point during the game. Frequently, you apply the rules of probability to decide on the best approach to winning the game. Games thus offer adults a way to practice mental math and give youngsters experience in applying number concepts. On top of all that learning, playing games with others builds strong bonds among friends and families. This chapter discusses a few games that develop math skills. They are listed in order of the age at which the game is first appropriate for play, youngest first.

When playing games, children also learn about taking turns and following directions, and they learn that they will not always win.

Games for Kids under 5

There are a variety of games that help children build their math and problem-solving skills—and they have fun in the process!

Candy Land

This board game by Hasbro is recommended for players ages 3 and up. The game board consists of squares that repeat in a 6-color pattern, and some of the squares have special characters on them. The play is determined from a set of cards, and each card may have one or two squares of a color on it. If there is one square, the player moves to the next square along the path with the corresponding color. If there are two squares of a given color, the player moves to the second square along the path of that color. If the card contains a special character, the player moves to the square with that character on it (this includes the possibility of moving backward).

Children who play Candy Land learn about patterns (e.g., blue squares come after yellow squares), and become able to move around the board more quickly as they gain experience. See *www.hasbro.com* for a complete set of directions for Candy Land.

Youngsters learn the art of counting in a very tactile way. They count their fingers and their toes. They count their toys. Playing board games continues to develop their understanding of numbers as they count the spaces they land on, or shape recognition, as they move to a square that matches the directions they are given. Playing games is an important part of the development of mathematical skills for young people.

Chutes and Ladders

This game by Hasbro is also recommended for players ages 3 and up. It is based on an old Indian game similar to Parcheesi. Mathematically, the children learn to recognize numbers and to count from 1 to 100. They learn some "skip counting" when they move up the ladder or fall down the chute. And they apply some basic probability when they start to estimate their chances of getting the number they need to reach a certain point on the board (such as the base of a ladder or the top of a chute) or the number they need to get to 100. Being able to answer a questions such as "What number comes after 8?" and "What number comes after 28?" leads to an understanding of how the number system works. When directed to "Show me a 5," the child should be able to hold up a hand with all 5 fingers displayed (or, if barefoot, point to the toes on one foot). See *www.hasbro .com* for a complete set of directions for Chutes and Ladders.

Feed the Kitty

This game, manufactured by Ceaco, is recommended for players ages 4 and up. Each player is given the same number of wooden mice, and play is based on the roll of two dice. The faces on each die contain not numbers but images (an arrow, a mouse, a sleeping cat, and a food dish). The children learn basic probability as they anticipate the faces that are rolled and the impact that each roll has on the number of mice in front of them. Although a player with no mice cannot roll the dice, such a player stays in the game, because it is possible that a player in an adjacent seat will be directed to pass a mouse to him or her. The game ends when one player has all the mice. See *www.howdoyouplayit.com* for a complete set of directions for Feed the Kitty.

Games for Children 6 and Up

As children grow, they require a different type of stimulation to learn math. Playing games such as Sorry and Uno can help them hone their skills and build their confidence in using math.

Sorry

This Parker Brothers games is recommended for players ages 6 and up. Play is controlled by a deck of cards that comes with the game. Each numbered card has a specific set of directions (these directions are written on the cards). For example, a card with the number 2 on it allows the player to take another turn. A card with a 4 on it requires that the player move backward, the directions that appear on a card with a 7 on it can be split between two of the game pieces, and a card with a 10 on it gives the player the option of going forward 10 spaces or backward 1 space. Strategy comes into play because there are slides that allow the player to move a number of extra spaces, and because any movement that ends with a player's game piece landing on the same square as an opponent's game piece sends the opponent's game piece back to Start (usually accompanied by a not-so-contrite "Sorry"). The child applies basic probability in determining strategy and counting to move game pieces. See *www.hasbro.com* for a complete set of directions for Sorry.

Uno

This Mattel card game is recommended for players ages 7 and up. The deck consists of cards numbered 1 through 9 in the colors red, blue, green, and yellow. There are also cards in these colors with the directions Skip, Reverse, and Draw Two. Finally, there are the black cards Wild and Draw Four. Players apply basic probability to develop strategy as they play (particularly when the next person in turn has just one card). When a player has discarded her or his last card and announces that she or he is "out," all the other players must count the points for the cards remaining in their hands. Numbered cards are worth face value; Skip, Reverse, and Draw Two cards are worth 20 points; and black cards are worth 50 points. This gives players young and old the chance to practice adding strategies (such as finding cards that add to 10), which makes it easier to count quickly and accurately. See *www.unorules.com* for a complete set of directions for playing Uno.

Games such as Uno, Chess, and Checkers require the player to concentrate on the current status of the game, anticipate the move of the next player, and to adapt to whatever moves are made before the player's next turn. These skills are essential to learning and using mathematics.

Shape by Shape

This game, by ThinkFun, is recommended for players 8 years old and older. Shape by Shape is a game of spatial relationships, as players use Tangram tiles to replicate one of the 60 two-dimensional patterns that come with the tiles. Children may also use the tiles outside of the playing of the game to make their own designs. See *www.puzzles.com* for a teacher's guide for Shape by Shape.

Block by Block

This game by ThinkFun is also recommended for players 8 and older. Block by Block involves the use of block pieces to reproduce designs that are provided. As in Shape by Shape, the players gain practice perceiving and working with spatial relationships—but this time in three dimensions.

Spatial relationships are as important as number relationships. Solving jigsaw and physical puzzles are an excellent way to develop these skills. The computer game Tetris is also an excellent way to develop spatial skills.

EqualZ

This card game from Karmel Games is recommended for players ages 10 and up. The deck contains numbered cards (0 through 9), the four basic operators (+, −, ×, and ÷), and parenthesis cards. The game includes two decks, for 90 cards. With 2 or 3 players, you use one deck. With 4, 5, or 6 players, you use both decks. The game features a very simple goal. At the beginning of each round, the dealer turns two number cards face-up, and these form the target number. Then each player tries to make mathematical equations that equal that target number. This is an excellent way for students to practice their computational skills and to grasp the importance of the order of operations in arithmetic.

For example, if the dealer turns over the cards 2 and 7 to make 27, a player may play 3×9, $20 + 7$, or possibly $(11 - 8) \times (5 + 4)$. You get 30 points for each expression you make, plus the point value of the number cards in the expression, plus 5 points for each operator. No points are awarded for parentheses. The player who gets rid of all his or her cards gets a 50-point bonus. Players who end the round with cards in their hand must subtract the point values of the number cards and 5 points for each operator left in their hand.

More Complex Games

Just as some games are made for children, some were created with adults in mind. Even though it may be difficult for children to take part, they will learn a lot just by trying. And with encouragement and practice, some of them may teach the grownups a thing or two!

Scrabble

This game by Hasbro has the advantage of developing vocabulary as well as math skills. Players need to determine what words they can make to get maximum points by taking advantage of double- and triple-letter squares, double- and triple-word squares, and the words formed in an adjacent row or column, and by avoiding the hazard of giving opponents "easy openings" for high-scoring words. Each player starts with seven letter-bearing tiles that opponents cannot see. Players take turns adding letters from their hidden supply to spell words on the board, replacing any letters that they use by drawing "blind" from the general supply of letters remaining. The object is to amass the most points by the time the game ends, which happens when all the tiles are in play but no player can make another word. Some cells on the board are marked to boost the value of placing a letter on them: DL for Double Letter, TL for Triple Letter, and DW for Double Word. For example, let's say the board looks as follows, and the next player has the letters T_1 S_1 E_1 K_5 O_1 A_1 D_2 to play with (the numbers in subscript represent the point values for the letters).

							DW	
DW						DW		
	TL			TL				
		DL	DL					
		G_2	R_1	A_1	I_1	L_1		
		DL	O_1	DL				
	TL	T_1	TL					
DW				DW				
							DW	
								DW

If the player puts down the word STEAK (adding to the word GRAIL), he or she scores points for the two words GRAILS ($2 + 1 + 1 + 1 + 1 + 1 = 7$) and STEAK ($1 + 1 + 1 + 1 + 5 = 9$), for a total of 16 points. But think again:

							DW	
DW					DW			
	TL				TL			
		DL		DL				
		G_2	R_1	A_1	I_1	L_1	S_1	
		DL	O_1	DL				T_1
	TL		T_1		TL			E_1
DW					DW	A_1		
							DW	K_5
								DW

The double-word square sitting directly below the K is just waiting for an opponent to add an S and score at least 20 points. Had the player put down the word STAKED (or STOKED or SKATED), the score would have been 7 points for GRAILS and $(1 + 1 + 1 + 5 + 1 + 2) \times 2 = 22$ for STAKED, for a total of 29 points.

								DW	
DW						DW			
	TL			TL					
		DL	DL						
			G_2	R_1	A_1	I_1	L_1	S_1	
		DL	O_1	DL				T_1	
	TL		T_1		TL			A_1	
DW						DW		K_5	
							DW	E_1	
								D_2	

If the letter had been such that the word created used both the Double Word scores that are 3 rows from I in GRAIL, the total value of the letters used would have been multiplied by 4. Using all his or her seven letters in one turn garners the player an additional 50 points. See *www.scrabblepages .com/scrabble/rules/* for a complete set of directions for playing Scrabble. A variation of Scrabble called Words with Friends is available through Facebook or as an app for a smartphone or iPod.

KenKen

Invented by a Japanese math teacher, Tetsuya Miyamoto, KENKEN® enables you to test your puzzle acumen and improve your math skills at the same time. The game is a variation of Sudoku®, the puzzle in which you must use each number from 1 through 9 once in each row, in each column, and in each square. But in KenKen, there are arithmetic problems to solve. For this 4 × 4 example, the numbers 1, 2, 3, and 4 are used to complete the problems.

```
┌──────┬──────┬──────┬──────┐
│ 11+  │      │ 4+   │      │
│      │      │      │      │
├──────┼──────┼──────┤      │
│      │ 24x  │      │      │
│      │      │      │      │
├──────┼──────┼──────┼──────┤
│ 2÷   │ 2    │      │ 1-   │
│      │      │      │      │
│      ├──────┼──────┤      │
│      │ 2-   │      │      │
│      │      │      │      │
└──────┴──────┴──────┴──────┘
```

The heavy boundary lines show the results of an arithmetic problem. The first two cells in the first row and the first cell in the second row make up a block where the three numbers in the block must add to 11. The remaining cells in the first column must have a quotient of 2 because the operator reads 2÷. In KenKen, order does not matter in division and subtraction. As a result, it could be that the number in the first cell of row 3 is a 2 and the number in the first cell of row 4 is a 1 (giving a quotient of 2) or it could be that the number in the first cell of row 3 is a 1 and the number in the first cell of row 4 is a 2. It could also be that the two numbers for the first cells in rows 3 and 4 are a 2 and a 4.

Note that the second cell in row 3 is a block unto itself with no arithmetic operator included. This tells you that the number 2 must be placed in the cell and that the number 2 cannot appear anywhere else in row 3 or column 2. The directions 24 × and 11+ are good clues to help you begin. Using only the numbers 1 through 4, which three numbers will multiply to be 24? The answer must be 2 × 3 × 4. The question is where to put them. You cannot have a 2 in row 3 or in column 2, so the 2 must appear in row 2, column 3. Where do the 3 and 4 go? The answer to that is based on the clue 11+. It is true that 1 + 2 + 3 + 4 = 10, not 11. How to get a sum of 11? The trick is that not all the cells are in a line (one row or one column), so numbers within the block can be repeated. You know that $4 + 3 + 4 = 11$, and the only way to enter these numbers into the block, while obeying the rule that the number must appear only once in each row and each column, is for the first two cells in row 1 to be a 3 and a 4, in that order, and for the first cell in row 2 to be a 4.

11+ 3	4	4+	
4	24x 2		
2÷	2 2		1-
	2-		

This makes it clear that the cell in row 2, column 2 must be a 3, and the cell in row 3, column 3 must be a 4 to complete the product of 24 block. The cells in column 2 now have all but one number filled in, so the last cell in column 2 must be a 1. The number in the cell formed by row 4, column 3 must be a 3 (so that the difference in this block of two cells is 2). This pretty much launches a domino effect for all the other cells. The number in the cell row 1, column 3 must be a 1 to complete column 3, and then the number in row 1, column 4 must be a 2 to complete row 1. The last number in row 2 must be a 1 (to complete row 2 and to be consistent, because $1 + 2 + 1 = 4$). Fill in the rest of the cells to get

11+ 3	4	4+ 1	2
4	24x 3	2	1
2÷ 1	2 2	4	1- 3
2	2- 1	3	4

KenKen appears in many newspapers and can also be played online. You will notice that the puzzles are not limited to four-number squares and that there are different levels of difficulty in the puzzles. Solving these puzzles is a wonderful way to practice arithmetic *and* logical thinking.

Cribbage

Cribbage is the card game that has been the basis for most of the rummy games. It is played with a pegboard and a standard deck of cards, and the goal is to reach 121 or more points first. The mathematical skills used in the game are addition, counting, and probability. The game begins with determining who will deal first (this is usually decided by cutting the deck, and the player with the lower card deals first). Each player is dealt 6 cards. Each player keeps 4 cards and deposits the remaining two cards into the "crib," a second hand that the dealer will use to count points after the play of the hand is completed. The person who did not deal the hand cuts the deck, and the top card is turned over (this is called the turn card). If the turn card is a Jack, the dealer receives 2 points for "his nibs," a words used to refer to a superior—and self-important—person in old England, where the game was created. If the card is not a Jack, the person who did not deal plays the first card. Assume that the turn card is the Q♥. As the player lays down the card in front of himself, the point value for the card is announced. The next player to play a card announces the sum of the point value of that card and the total before that card was played; thus each player adds to the running total. For example, if player 1 plays 4♣, he announces "4." The dealer plays 8♥ and announces "12." Player 1 plays 3♦ and announces "15."

There are a number of ways in which scoring occurs. Reaching the sum of 15 earns a player 2 points. In the example given, player 1 will score 2 points and will therefore move the peg on his board 2 places from the starting position. Points also can be pegged for pairs (or triples and quadruples), for runs of three cards or more, for reaching a sum of 31, and for playing the last card that does not cause the sum to exceed 31.

A run is a set of cards that are consecutive. For example, 2–4–3 is a run. (The cards do not have to be played in sequential order; any three that would be consecutive if rearranged form a run.) Ace is always low card and counts as a 1.

To continue the example of the play, the dealer plays the card 3♥, announces "18, pair for 2," and moves his peg 2 places on the pegboard. Player 1 plays 2♠ and announces "20." The dealer plays A♥, announces "21, run for 3," and moves his second peg 3 places in front of his first peg to the fifth hole. (The alternating use of pegs helps keep track of where you started to count from and how many places you've counted. The use of multiple pegs allows the player(s) to check the number of points they have counted in case there is an error.) Player 1 plays Q♦, announces "31, 2 points," and uses his second peg to move 2 places forward on the pegboard. Then the dealer plays his last card, 5♥, announces "5, 1 point for last card," and moves the first peg 1 space in front of his second peg on the pegboard.

The next phase of the hand is to count the points in the hand that was kept in the crib. (This is why the cards are laid down in front of the player, rather than in one pile generated by both players.)

Player 1 had the cards 4♣ 3♦ 2♠ Q♦, and the turn card was Q♥. Player 1 has a pair of Queens (2 points), a run 2–3–4 (3 points), and a sum of 15 twice with 3♦ 2♠ Q♦ and 3♦ 2♠ Q♥ (4 points), for a total of 9 points. Player 1 moves the back peg 9 places past his forward peg on the pegboard.

The dealer had the cards 8♥ 3♥ A♥ 5♥, and the turn card Q♥. The dealer has a four-card flush in his hand (good for 4 points), as well as the bonus that the turn card is also a heart (for an extra point). The dealer also has a sum of 15 with the Queen and 5, for 2 points, so the dealer has a total of 5 points. He moves his back peg 5 places in front of his forward peg.

The 8 cards just counted are set aside, and the dealer turns over the cards in the crib and includes the turn card. The cards in the crib are 6♦ 5♣ 4♦ 6♠, and the turn card is Q♥. The dealer has a double run (6♦ 5♣ 4♦ and 5♣ 4♦ 6♠) worth 6 points and a pair of sixes worth 2 points. Each run adds to 15, and the Queen-and-5 combination is 15, for 6 additional points. The points for the crib total 14.

FACT

A double run will always be worth 8 points because of the two runs and the one pair.

At this point in the game, player 1 has 13 points and the dealer has 22 points. The next hand continues with player 1 becoming the dealer.

In this hand, the pair of threes, the pair of Queens, and the pair of sixes each scored 2 points for the player. If one player had a triple (e.g., 8♠ 8♥ 8♦), the point value would be 6 points, because there are three different pairs that can be made with these cards (8♠ 8♥; 8♥ 8♦; 8♠8♦). Do you see how you can use the counting principle to show that the hand 8♠ 8♥ 8♦ 8♣ is worth 12 points because there are six different pairs that can be formed?

See either *www.cribbage.org/rules/rule1.asp* or *www.crosscribb.com* for a complete set of directions and rules for cribbage.

Games Online

Many of the games described in this chapter can be played online or on a personal computer, smartphone, or other electronic device. When you play in this format, you get some practice playing the game, but many of the arithmetic computations that you might want to practice are done for you. Yahoo has an extensive list of games (including cribbage, Sudoku, and jigsaw puzzles) that you can use to practice numerical and spatial skills. There is a free app for your iPod to play cribbage (one of the settings enables you to count the points at the end of the hand). There are also a number of sites for children to use to practice math skills. The site Math Games (*www .maths-games.org/counting-games.html*) has a number of activities for students of all ages. The site Fractions Mistakes (*www.tes.co.uk/sen-teaching-resources*) is from England and has work with fractions and decimals for children of middle-school age.

QUESTION

How do you use math when you play games?
Consider the games of chance and games of strategy that you enjoy. What else did you add to your list of when you use math?

Exercises

Find the value for each of the following cribbage hands.

1. 4♠ 5♠ 6♠ J♠, Turn card Q♦
2. 10♣ 10♦ J♦ Q♠, Turn card 5♦
3. 7♣ 7♦ 7♥ A♥, Turn card 8♥

Complete the Ken-Ken boards.

4.

12x		3x	4÷
	4		
2-		2÷	
3-		5+	

5.

4+		2÷	
3-		8+	
8+		1	
	12x		

Exercise Solutions

Chapter 2

1. 2,200
2. 17,800
3. 2,400
4. 300 (when rounding both of the numbers down or both up)
5. 700
6. 17,000
7. 40,000
8. 6,000 (when rounding both of the numbers down)
9. 1,238
10. 1,862
11. 295
12. 2,067
13. 713
14. 9,025
15. 1,577
16. 2,021
17. 1,075
18. 70
19. −520
20. 160
21. −624
22. 266
23. −20
24. 34
25. −205

Chapter 3

1. 6.5×10^7
2. 2.54×10^6
3. 5,865,696,000,000 miles; 5.87×10^{12} miles
4. 14,887,136,448,000,000,000 miles; 1.49×10^{19} miles
5. 94°F

6. −40°F (This is the one and only temperature that is the same on both scales.)
7. 27.5°C
8. $50 \times 0.005 = 5 \times 10 \times 5 \times 10^{-3} = 25 \times 10^{-2} = 0.25$
9. $50 \div 0.005 = 50 \times 1{,}000 \div 0.005 \times 1000 = 50{,}000 \div 5 = 10{,}000$
10. $15.39; the customer may have wanted a ten-dollar bill and a five-dollar bill rather than more singles.

Chapter 4

1. a, c, d

2. $\dfrac{35}{24} = 1\dfrac{11}{24}$

3. $\dfrac{121}{504}$

4. $\dfrac{16}{36} = \dfrac{4}{9}$

5. $4\dfrac{23}{24}$

6. $\dfrac{35}{96}$

7. $36\dfrac{11}{24}$

8. $\dfrac{14}{15}$

9. 0.5625

10. $\dfrac{9}{20}$

11. $\dfrac{5}{4}$-inch wrench

Chapter 5

1. $\dfrac{4}{5}$

2. 96

3. 2683.2

4. 16.4%

5. 133.3%

6. 86.7%

7. 975

8. 13.1%

9. $604.87 million

Chapter 6

1. $13.74
2. $32.35
3. The first bed is cheaper; $34.31
4. $27.15

5. $6
6. The 21-ounce box, at $4.66 per pound
7. $53.20

Chapter 7

1. $1,860
2. 1.5%
3. 0.666666667%
4. $2,500(1.015)^{20} = $3,367.14

5. $2,500 $\left(1 + \dfrac{0.08}{12}\right)^{60}$ = $3,724.61; a savings of $357.48

6. $32.50
7. $73.80

Chapter 8

1. 1 mL

2. $\dfrac{3}{4}$ cup

3. 12 (3 tsp. = $\dfrac{1}{16}$ cup, so $\dfrac{1}{2}$ cup = $\dfrac{8}{16}$ cup = 24 tsp.)

4. $1\dfrac{3}{4}$ cup

5. $69\dfrac{1}{3}$ fl. oz. (1 lb. = 16 fl. oz.; 4 lb. = 64 fl. oz.; $5\dfrac{1}{3}$ oz. = $5\dfrac{1}{3}$ fl. oz.)

Chapter 9

1. $2,013.85
2. $1,165.50
3. $2,027.08

Chapter 10

1. Approximately 3 hours 15 minutes
2. $51
3. $603.40 (less whatever they pay to eat along the way)

Chapter 11

1. In the neighborhood of $25,000 – $28,000 (with maximum suggested fair market value at $26,500)
2. (a) $586, (b) $724, (c) $1,015
3. $30,195; Sumit should sell the car himself for $31,925 and collect an extra $1,730.

Chapter 12

1. $200,000
2. $966.40
3. $181,186.18
4. $15,000
5. $53,813.82
6. $256,186.18
7. $1,409.46

Chapter 13

1. They will each have $10,583.98 in their accounts. (PV = 0, PMT = 120, r = 0.014, m = 12, t = 7)
2. $93,122.05 (PV = 10,583.98, PMT = 400, t = 15)
3. $222,958.24 (PV = 93,122.05, PMT = 540)
4. $1762.51 (FV = 0, PV = 445,916.48, t = 25)

Chapter 14

1. Line 4: 49,218.64; Line 5: 9,750; Line 6: 39,468.64; Line 7: 8,358.47; Line 8: -0-; Line 9: 8,358.47; Line 10: 5,899; Line 11: $2,459.47 Refund!
2. Line 4: 90,103.64; Line 5: 9,750; Line 6: 80,353.64; Line 7: 15,296.35; Line 8: -0-; Line 9: 15,296.35; Line 10: 16,124; Line 12: $827.65 is owed to the government.

Chapter 15

1. $\dfrac{15}{36} = \dfrac{5}{12}$

2. $\dfrac{132}{2652} = \dfrac{11}{221}$

3. $\dfrac{6}{54} \times \dfrac{5}{53} \times \dfrac{4}{52} \times \dfrac{3}{51} \times \dfrac{2}{50} \times \dfrac{1}{49} = \dfrac{720}{18,595,558,800} = \dfrac{1}{25,827,165}$

4. Mean: $6,135,906; Median: $2,425,000

5. Mean: $\dfrac{1(40) + 2(32) + 3(40) + 4(28) + 5(19) + 7(10)}{40 + 32 + 40 + 28 + 19 + 10} = 2.96$

Median: There are 169 pieces of data (the sum of the frequency column). The median will be the 85th piece of data (84 below and 84 above). There are 40 cases of 1 hour and 32 cases of 2 hours of overtime. This totals 72 pieces of data. The 85th piece of data will be in the group of 3 hours of overtime.

Chapter 16

1. 25 cm
2. Use the circumference formula $C = 2\pi r$ and solve for r. The radius is $r = 3.183$ in. Use the volume formula $V = \pi r^2 h$ and solve for V. The volume is approximately 382 in^3.
3. 1530 in^3
4. To be on the safe side, dimensions are rounded down so that the area remaining is larger than the "real" number. Area of small rectangular windows: $2 \times 3 \times 3.5 = 21$; area of fireplace: $4 \times 4 = 16$; area of rectangle in Norman window: $4 \times 4 = 16$; area of semicircle in Norman window: $\frac{1}{2} \times \pi \times 2^2 = 12.5$; area of quarter-circles: $2 \times \frac{1}{2} \times \pi \times 2.5^2 = 19$. Thus the entire painting area is $360 - 21 - 16 - 16 - 12.5 - 19 = 275.5$ sq. ft.
5. 3 inches above the top of the window (for a 1-inch gap at the floor); the brackets will be $25\frac{3}{4}$ inches from the center point on the window (or $\frac{1}{2}$ inch from the outside of the window frame).
6. $8.69
7. 5,200 sq. ft.
8. $24\frac{2}{3}$ ft.

Chapter 18

1. 13
2. 16
3. 18

4.

12x		3x	4÷
3	2	1	4
2	4 4	3	1
2- 1	3	2÷ 4	2
3- 4	1	5+ 2	3

5.

4+		2÷	
3	1	2	4
3- 1	4	8+ 3	2
8+ 4	2	1 1	3
2	12x 3	4	1

In the Craft Room:
Measure Twice, Cut Once

You may not consider yourself the "crafty" type, but every once in a while, you'll find yourself in a situation where you need to use math to create something. Maybe you're only interested in hemming some pants. Perhaps you want to take up scrapbooking. Maybe you want to learn to make new candles from the stubs you have stuck in your utility drawer. Whatever your goals, math sneaks into most craft projects.

Knitting and crocheting have always had a strong following. The geometry of yarn work is pretty simple, especially if you're making a scarf or a blanket. (Rectangles are simple.) Every good knitter or crocheter has a few easy math skills at the tips of her needles and hooks.

Here is a little anecdote to illustrate:

Last week, Ann found the most gorgeous yarn on sale at her favorite yarn shop, and she has just the project for it—a simple crocheted scarf. The problem is that the sale wasn't really that wonderful. She could afford only 4 skeins of the yarn, and her pattern calls for 6 skeins. No matter. She can make the scarf smaller for her daughter, Jasmine, who just turned 5. Ann decides to leave the width of the scarf the same and reduce the length. But how can she change the pattern? Her best bet is to do a little math.

Proportions are just the thing in this situation. They are ideal for showing the relationships among four numbers. (A ratio is a way to compare two numbers; proportions are just ratios on steroids.)

Her pattern tells Ann the relationship between the number of skeins and the length of the finished scarf. (She's already checked her gauge, so she knows she can trust the size.) The pattern calls for 6 skeins of yarn, and the finished scarf will be 66 inches long.

Ann writes that information as a ratio:

$$\frac{6}{66}$$

She has 4 skeins of yarn, but she doesn't know how long her scarf should be. Therefore, she creates another ratio to represent this information, using s as her stand-in for the length of the shortened scarf:

$$\frac{4}{s}$$

Now she's ready to create the proportion, and it is

$$\frac{6}{66} = \frac{4}{s}$$

Ann looks carefully at the equation she's created. Does it really express an equality? She needs the numerators to be "like items" (the numbers of skeins) and the denominators to be "like items" (the lengths of the scarves). Satisfied, she moves on.

To solve the proportion, Ann just needs to cross-multiply. She'll have to multiply the numerator of the first ratio by the denominator of the second ratio and the numerator of the second ratio by the denominator of the first ratio. (That's a lot easier *done* than *said*, actually!)

$$\frac{6}{66} = \frac{4}{s}$$

$$6s = 4 \cdot 66$$

$$6s = 264$$

Now she can solve for x by dividing each side of the equation by 6.

$$\frac{6s}{6} = \frac{264}{6}$$

$$s = 44 \text{ inches}$$

Aha! Her finished scarf will be 44 inches long. Now that the math is out of the way, Ann can get to work.

ESSENTIAL

Open up any knitting or crochet book, and you'll be instructed to make a gauge or tension swatch to make sure that your finished piece will be the right size. If you crochet too tightly, the scarf you make might not be long enough to wrap around your boyfriend's neck. If you knit too loosely, a medium-size sweater might fit a linebacker. So, before starting a project, knit or crochet a sample, using the same yarn and the same needle or hook size. The pattern will tell you how long a certain number of stitches should be. For example, a 10-cm knitted square might need to be 22 stitches and 30 rows, using 4-mm needles. When you're finished, compare your sample with the gauge on the pattern. Too small? Up your needle or hook size. Too big? Go down a size or two.

Math? Sew What!

Everyone needs to know how to sew at some point in their life—maybe just a button, the hem of a dress, a tear in a sleeve. But how do we know how much thread to use? Consider Reggie. Reggie is a newcomer to sewing.

After finding his mother's sewing machine, he looks around for his first project. What's a really easy project to start out with? Pillow covers.

Doing the Calculations

First, Reggie must figure out how much fabric he'll need. He hunts for his new tape measure. The pillows are indeed square, measuring 1½' by 1½'. He pulls out a piece of paper and sketches a simple square. Then he labels his drawing.

Easy enough. For each pillow, he needs two pieces of fabric measuring 1½' by 1½'—one piece for the back and one piece for the front. Because he is making two pillows, he'll need four pieces all together.

Then, Reggie remembers seam allowances. If he's going to sew the fabric together, he needs to make room for the seams. But how much room? On a website for beginning sewers, he learns that seam allowances are typically ⅝". He looks again at his sketch, thinking about where the seams will go.

Reggie knows there are four seams in each of his pillows—one for each side—so he needs to add ⅝" to each side of the square. Doing some calculations, this is what he comes up with:

Front and back pieces of each pillow = ⅝" + 1½' + ⅝"

Reggie notices two things right away: Not only is he dealing with mixed numbers *and* fractions, but he's also got two different units of measurement: inches and feet. He's going to have to do some conversions. He decides on inches.

$$1½' = 1' + ½'$$
$$1' = 12"$$
$$½' = 12" \bullet ½, \text{ or } 6"$$
$$1½' = 12" + 6"$$
$$1½' = 18"$$

(You say there's an easier way to do this? Yes, you can simply multiply 1.5 by 12. You'll get the same answer. But know this: Fractions are huge in sewing. Getting some practice with them is not a bad idea if you have sewing aspirations yourself.)

Reggie revisits his addition problem, this time using 18" in place of 1½'. And now that everything is in the same unit of measurement, he doesn't need to include those annoying little unit marks.

$$\frac{5}{8} + 18 + \frac{5}{8}$$

Clearly, Reggie still has work to do. He can get a common denominator for all of the fractions. Or he can try a different process, to make things easier on himself.

Reggie had the problem arranged this way because he was picturing his pillow: seam allowance+size of pillow+seam allowance. But this is addition, so he doesn't *have* to add in that order. It makes more sense for him to add the fractions together first and then see what he can do.

$$\frac{5}{8} + \frac{5}{8} + 18$$

$$\frac{10}{8} + 18$$

Now he's faced with another choice. $\frac{10}{8}$ is an improper fraction—in other words, it's larger than 1. Should he change it to a mixed number—a whole number and a fraction? He decides to do so. That's because it's pretty straightforward to add a mixed number to a whole number.

To change an improper fraction to a mixed number, just divide the numerator by the denominator. The answer is your whole number, and the remainder is the numerator of the fraction part. The denominator stays the same.

He scribbles these steps in the margin.

$$\frac{10}{8} = 1\frac{2}{8}$$

But wait. There's something up with Reggie's answer. The fraction of the mixed number isn't in its simplest form. That's not necessarily a huge deal, but this fraction is easy to reduce. All he needs to do is find a common factor for both the numerator and the denominator and then divide each by that number. The 1 just comes along for the ride.

$$1\frac{2}{8} = 1\frac{2 \div 2}{8 \div 2} = 1\frac{1}{4}$$

Thus Reggie's seam allowances add up to 1¼". He can finish the problem now:

1¼ inches + 18 inches = 19¼ inches

That means Reggie needs four squares of fabric that measure 19¼ inches on all four sides.

Putting the Calculations to Use

Buying the fabric is the next step. Reggie heads to the decorator fabrics, looking for something sturdy, without a lot of stretch. Two fabrics catch his eye: a lime-green polka dot on cream and a fuchsia paisley. The polka-dot fabric is 54" wide, and the paisley is 48" wide. This is where his calculator will come in handy.

How many pillows can Reggie fit along the width of the fabric?

Each pillow is 19¼" wide. He knows he needs to divide the width of the fabric by that number. Instead of trying to convert the ¼" to a decimal, Reggie just rounds down to 19. Dealing with the 54"-wide fabric first, he divides:

54" ÷ 19" = 2.8
Now the 48"-wide fabric:
48" ÷ 19" = 2.5

Reggie can get only two pieces across each of the fabrics. Which one should he buy? He'll have less fabric left over with the second one, and besides, he always thought that room could use a punch of pink. Fuchsia it is!

But how many yards? The length of the side of his sketch will help here, too. He needs pieces for two pillows. Because two pieces will fit on the 19" length of 48"-wide fabric, he needs just two 19" lengths (19" • 2 = 38").

However, fabric is sold by the yard, so he still needs to convert again. How many yards is 38"?

There are 36 inches in 1 yard (12 inches in a foot and 3 feet in a yard: 3 • 12 = 36). Reggie needs to divide the number of inches he needs by 36 to find out how many yards he needs.

39 inches ÷ 36 inches = 1.09 yards or 1 yard 3 inches

He needs a little more than 1 yard, but not much. Reggie decides on 1¼ yards. Now all he needs is a spool of thread, some needles, and a few hours of free time.

Metric Units

You may find yourself dealing with metric units occasionally. This chart will help you convert metric units to standard units.

1 in.	25.4 mm	2.5 cm
1 ft.	0.305 m	30.5 cm
1 yd.	0.914 m	91.4 cm

Knowing some basic fraction-to-decimal conversions is really helpful, too.

⅞	0.875
⅘	0.8
¾	0.75
⅝	0.625
⅗	0.6
½	0.5
⅖	0.4
⅜	0.375
¼	0.25
⅕	0.2
⅛	0.125

Side to Side

Fabrics have standard widths that vary by type of fabric. Here's a typical list of fabric widths, although they may vary from manufacturer to manufacturer.

Type of Fabric	Possible Widths
Decorator fabric	48" and 54"
Fashion fabric	36", 45", and 60"
Quilting fabric	42" to 44"

If you're lucky enough to score vintage fabrics, you might find some even stranger widths. By the way, because width is measured in inches, use inches when you convert your measurements. It's much easier to figure out what you need if you're already working with like units.

Math and Scrapbooking

If you are thinking about taking up scrapbooking, you'll need to have a few basic math skills in place to help you get the job done. Consider Mary—she's a terrific photographer, and she'd like to highlight her art in some special ways. Any new hobby requires materials, and scrapbooking is no exception. Mary picks up an album, some acid-free paper, adhesives, and a few nice pens. Then she starts thinking about the types of pages she wants to build.

Mary loves spreadsheets, so she opens one up and starts typing. She read somewhere that a two-page spread shouldn't have more than 10 photos. She can break this rule if she wants, but it sounds like a good frame of reference.

Mary starts categorizing her photos, and here's what she ends up with:

Categories	Number of Photos	Number of Pages
The kids: Ella	20	
The kids: Mabel	15	
The kids: Josh	18	
Halloween	10	
Beach vacations	48	
Holidays with the folks	32	
Road trip out West	12	

Clearly, she's going to need more pages for some subjects than for others. There's something else Mary needs to consider. Her final single-page count must be divisible by 4. That's so she doesn't have any blank pages. Using trial and error, Mary finagles the number of photos and the number of pages until she comes up with this count.

Categories	Number Of Photos	Number Of Pages
The kids: Ella	20	4
The kids: Mabel	15	4
The kids: Josh	18	4
Halloween	10	2
Beach vacations	48	6
Holidays with the folks	32	6
Road trip out West	12	4

Using her spreadsheet application to total the pages, she finds that she's planned for 30 pages. But that's a problem. Thirty is not a multiple of 4. In other words, 4 doesn't divide evenly into 30.

Mary thinks a moment. What multiple of 4 is closest to 30? Well, 28 and 32 are both close to 30. Should she add 2 pages or subtract 2 pages?

Mary thinks again. Then she picks up her copy of *Moby Dick* from the coffee table. There's the cover—just like the cover of her album—and the title page, then hundreds of 2-page spreads, and finally the back page. Mary hadn't considered the first and last pages!

It makes sense for her to add 2 pages. That way, she can do a nice title page and wrap things up at the end—maybe with a cute picture of Grandma and the kids.

She has a plan. Now all Mary needs to do is figure out the differences between glues, adhesive tapes, and photo corners.

FACT

A *multiple* (also called a product) is what you get when you multiply two numbers. Any multiple of a number is evenly divisible by that number—there will be no remainder. So, if you multiply 9 by 3 to get 27, the multiple (27) is divisible by that number (9); there are no fractions left over. Here's another example, $4 \cdot 6 = 24$. Twenty-four is a multiple of 4 and 6, because you multiplied 4 and 6 to get that multiple. The multiple (24) is evenly divisible by 4 (the quotient, or answer, is 6) and also is evenly divisible by 6 (in which case the quotient is 4). But wait! There's more! A *factor* is a number that divides evenly into another number, so 4 and 6 are factors of 24.

Math and Pictures

If your photos are stored electronically, you can size them easily using software. But before that software existed, book publishers, newspaper designers, and even yearbook editors had to depend on good old proportions to blow up or shrink photos. And in fact, this is exactly how image software works today.

Remember, a proportion is a pair of equal ratios. If you change one of the numerators, you have to change its corresponding denominator in the same way. Otherwise, the ratios will no longer be equal.

Let's say you have a photograph that is 6 inches tall by 4 inches wide. You'd like to enlarge it so that the width is 6 inches. If you change the width without changing the height, you'll have an odd-looking photo. So if you change the height proportionally to the change in the width, what will the height of the new photograph be? First, set up the proportion, using the heights as the numerators and the widths as the denominators.

$$\frac{6}{4} = \frac{x}{6}$$

Then cross-multiply, and solve for x.

$36 = 4x$

$9 = x$

The picture must be 9 inches tall. What if you set up the proportion with the widths as the numerators and the heights as the denominators?

$$\frac{4}{6} = \frac{6}{x}$$

$4x = 36$

$x = 9$

How about that—the same answer. Math *is* flexible!

Economies of Crafting

Before pulling out the sewing machine or setting up the table saw, ask yourself, "Does it make more sense to buy what I'm about to make?" You may want the experience of building something from scratch. But you also may need to keep a little change in your pocket. It's always a good idea to consider what your craft will cost the family budget.

For example, Rita loves Halloween, and she loves making her kids' costumes. This year, her 10-year-old daughter has requested a velvet-like cape and gown so that she can dress as some obscure character from her favorite novel about magical kids.

The pattern Rita is using calls for 7 yards of fabric, 2 fancy fasteners, and 3 yards of fringe. Looking at the Sunday circular for the local fabric store, she sees that crushed panne velvet is on sale for $2.99 per yard and the fringe is priced at $4 per yard. Rita guesses that the fasteners are about $5 each. To estimate her costs, she adds everything together:

$$(7 \cdot \$2.99) + (3 \cdot \$4) + (2 \cdot \$5)$$

(In case you lost track, that's 7 yards of fabric at $2.99 per yard, 3 yards of fringe at $4 per yard, and 2 frog clasps at $5 each.)

$$\$20.93 + \$12 + \$10 = \$42.93$$

A terrifying price! Rita is starting to think that a trip to a thrift shop might be a better investment of her time and money. Sometimes, doing it yourself just isn't worth it.

Don't Complicate Things

On math worksheets from school, you have exactly the information you need. In real life, though, you have to decide which numbers are relevant—and how. That's not always as simple as it sounds.

For example, Tad wants to build a birdhouse out of scrap wood. Here is what he knows:

Base of the interior: 4" by 4"
Height of the interior: 10"
Distance of entrance from the floor: 8"
Diameter of entrance: 1⅛"
Height above ground: 6' to 15'

Which of the pieces of information on this list will he use to build the birdhouse? Basically, Tad will be constructing a box with a big hole drilled into it. To figure out how much wood he'll need, he uses the dimensions of the box: 4" by 4" by 10". Those dimensions tell him the dimensions of each side.

Each of 4 sides: 4" by 10"
Top and base: 4" by 4"

Tad will need to cut four pieces of wood that measure 4" by 10" and two pieces that measure 4" by 4". (Any box has six sides.) The next two measurements in his list tell Tad where and how large to drill the entrance to the birdhouse.

But that last bit of information? It's not relevant. It is only *after* he's built the birdhouse that Tad needs to know how high to install it.

APPENDIX C

Math in the Gym

It's time to hit the gym. You probably already know that maintaining weight is a balancing act: Calories in = calories out. To lose weight, you'll need to shake things up: Fewer calories in + more calories out = weight loss. Yep, there's math involved in losing weight and staying fit. Whether you're at the gym or the kitchen table, a few computations can keep you on the right track.

Your Ideal Weight

There are hundreds of different criteria for tracking changes in your weight—how you look in your favorite pair of jeans, whether you can still play touch football without getting out of breath, etc. However, for nutritionists, physicians, and personal trainers, that number on the bathroom scale matters. The pros also count on another important number: BMI, or *body mass index*. BMI is used to evaluate a person's health on the basis of their body weight and height. Here's the formula:

$$BMI = \frac{703w}{h^2}$$

w is weight in pounds

h is height in inches

Let's say June weighs 155 pounds, and is 5 feet 2 inches tall. What is her BMI?

Before you start substituting the numbers into the formula, take a look at what you have. To use the BMI formula, you need to know June's weight (in pounds) and her height (in inches):

$w = 155$ pounds

$h = 5$ feet 2 inches

But June's height is listed in feet *and* inches. That number needs to be converted to inches only. There are 12 inches in a foot, so multiply 5 feet by 12 and add the leftover 2 inches, like this:

$(5 \bullet 12) + 2 = 62$ inches

Now you can use the BMI formula:

$$BMI = \frac{703w}{h^2}$$

$$BMI = \frac{703 \bullet 155}{62^2}$$

$$BMI = \frac{108,965}{3,844}$$

$BMI = 28.35$

June's BMI is 28.35. So what?

Here's a chart (this one comes from the World Health Organization) to help you figure things out:

Classification	BMI
Underweight	<18.50
Severe thinness	<16.00
Moderate thinness	16.00–16.99
Mild thinness	17.00–18.49
Normal range	18.50–24.99
Overweight	≥25.00
Pre-obese	25.00–29.99
Obese	≥30.00
Obese class I	30.00–34.99
Obese class II	35.00–39.99
Obese class III	≥40.00

Where does June fit on the BMI table? She's considered overweight and pre-obese. Her doctor should suggest that she lose a few pounds to get into the normal range.

QUESTION

What are those funny looking symbols in the BMI table?
The sign opens up in the direction of the larger number. The big side of the symbol corresponds to the bigger number. "Greater than or equal to" (≥) and "less than or equal to" (≤) are the hybrids. So if your BMI is 25, are you considered overweight? The table reveals that the answer is yes. That's because when a person's BMI is greater than or equal to 25 (BMI ≥ 25), he or she falls in the overweight category. Now, get on that treadmill.

Eat Less, Move More

Everybody knows someone who can eat anything and stay slim. It may not be fair, but each of us burns calories differently. Some folks have good genes—they run through calories like water, which means they can eat what they want and forgo long sessions at the gym. Others seem to gain a pound by simply looking at the leftover French fries.

Fair or not, gender, age, weight, and height all play a role in how efficiently your body handles energy or calories. Did you notice? Many of these variables are numbers. Your age: a number. Your weight: a number. Your height: another number.

And where there are numbers, math is bound to be right around the corner.

Figuring Out Your Basal Metabolic Rate

Your *basal metabolic rate* (BMR) is another important number. It describes how many calories you would need to stay alive if you were to spend all day in bed asleep. In other words, BMR is the minimum calorie intake for a resting individual.

Of course, BMR varies from person to person. If you know your BMR, you can calculate the number of calories you need to consume in a day. But what is BMR based on?

A person's size is important: The more a person weighs, the more energy it takes to do daily tasks. This additional energy translates into a higher BMR. And as you age, you tend to burn calories less efficiently. That's because muscle mass decreases with age—particularly when achy joints and busy schedules keep us from exercising as frequently. It also means that, in order to maintain your weight as you age, you may need to take in fewer calories—or, better yet, burn more calories.

Finally, because of their muscle mass, men need to consume more calories than women, so gender plays a role. So, there are two formulas for BMR, one for women and one for men.

For Women:

$655 + (4.3 \bullet \text{weight in pounds}) + (4.7 \bullet \text{height in inches})$
$- (4.7 \bullet \text{age in years})$

For Men:

$66 + (6.3 \bullet \text{weight in pounds}) + (12.9 \bullet \text{height in inches})$
$- (6.8 \bullet \text{age in years})$

Want to get really geeky? Here are the formulas using variables:

$BMR_{women} = 655 + 4.3w + 4.7h - 4.7a$
$BMR_{men} = 66 + 6.3w + 12.9h - 6.8a$

w is weight in pounds

h is height in inches

a is age in years

These formulas show that BMR depends on gender, weight, height, and age. In other words, your gender, size, and age play a role in how efficiently you burn calories.

Let's look at an example. Let's say your best friend, Susan, weighs 185 pounds, is 5 feet 7 inches (67 inches) tall, and is 25 years old. What is her BMR? To find out, you'll use the formula for women, and substitute the information that you have for Susan:

$655 + (4.3 \cdot \text{weight in pounds}) + (4.7 \cdot \text{height in inches})$

$- (4.7 \cdot \text{age in years})$

$655 + (4.3 \cdot 185) + (4.7 \cdot 67) - (4.7 \cdot 25)$

$655 + 795.5 + 314.9 - 117.5$

1,647.9 calories

Adding in Activity

But remember, BMR tells you how many calories your body will burn if you were asleep all day. And that doesn't require much energy at all. Even climbing out of bed, pouring a cup of coffee, and brushing your teeth burns calories. So to find the number of calories you can burn in a day without gaining weight, you'll need to do one more calculation, and this one is based on your activity level.

Total calories $= \text{BMR} + (\text{BMR} \cdot \text{level of activity})$

Your level of activity is represented as a percent:

- Sedentary 20%
- Lightly active 30%
- Moderately active (exercise most days a week) 40%
- Very active (exercise intensely and daily or for prolonged periods) 50%
- Extra active (hard labor or athletic training) 60%

If you're sedentary, your recommended daily calorie intake would be your BMR plus 20% of your BMR. As an equation, that is

Total calories $= \text{BMR} + (\text{BMR} \cdot 0.20)$

(Did you catch that? You have to turn the percent into a decimal before you can multiply. All you need to do is move the decimal point two places to the left, and then drop the percent sign: 20% = 0.20.)

Let's take a look at Susan again. Her BMR is 1,647.9. She walks a couple of miles two or three days a week, so she's lightly active. So this is how she'll find the total number of calories she should consume each day to maintain her weight:

BMR + (BMR • 0.30)

1,647.9 + (1,647.9 • 0.30)

1,647.9 + 494.37

2,142.27 calories

(Do you see why she used 0.30, rather than the 0.20 that a sedentary person would use?) In order to maintain her weight, Susan must take in 2,142.27 calories each day.

Fat Chance

A pound of body fat equals just about 3,500 calories. This means that in order to lose 1 pound of fat, you need to consume 3,500 fewer calories (or burn 3,500 more calories).

However, if you look at those weight-loss advertisements, they make promises like "Lose at least 7 pounds a week on MeltAway, a revolutionary new diet pill that melts fat!" Is that reasonable?

Take a look at the math. To lose 1 pound, you need to reduce your calorie intake by 3,500 (or burn an extra 3,500 calories a day). If you were to lose 7 pounds of fat in a week, you'd be reducing your calorie intake by an average of 3,500 calories every day. Now remember that an average man needs 2,500 calories a day and an average woman needs 2,000. Unless you're consuming an extra 3,500 calories a day, you can't reduce your calories by that much.

But could you burn that many calories a day? Sure, and here's how: 7 hours of high-impact aerobics, 6 hours of mountain biking, 8 hours of golf (carrying your own clubs), or 13 hours of weightlifting. Oh, and those totals? They're daily, not weekly. So, unless those pills are revving up your body, hummingbird-style, if you do lose 7 pounds in a week, it won't be through loss of fat.

FACT

During the 2008 Summer Olympics, swimmer Michael Phelps made big news—not only for his impressive collection of gold medals, but also for his eating habits. Rather than the typical 2,500 to 3,000 calories a day that most men need to maintain their weight, Phelps consumed a staggering 12,000 calories each day.

Fast and Easy

There simply is no quick or simple way to lose weight, unless you're resorting to unhealthful methods. But there is an easy way to estimate your daily calorie intake. This method isn't as exact as the formula presented earlier, but it works great for those who aren't interested in precision.

- To lose fat, eat 12 to 13 calories per pound of bodyweight
- To maintain weight, eat 15 to 16 calories per pound of bodyweight
- To gain fat, eat 18 to 19 calories per pound of bodyweight

If you weigh 155 pounds, here's your calorie intake for each scenario:

Lose weight	12•155 to 13•155—that is, 1,860 to 2,015—calories per day
Maintain weight	15•155 to 16•155—that is, 2,325 to 2,480—calories per day
Gain weight	18•155 to 19•155—that is, 2,790 to 2,945—calories per day

(Note that these guidelines are the same for men and women. The resulting calories per day might be on the high end for women and on the low end for men.)

Many nutritionists also recommend looking at an average daily calorie intake over a week. If you weigh 155 pounds and average 1,900 calories a day for a week, you're likely to lose weight. If you average 2,800 calories a day for a week, the needle on the scale will probably move up.

A Well-Balanced Nutrition Label

A product's nutrition label is made up of several distinct parts, and each part is designed to help you make good choices about what you're eating. But which numbers are important, and why?

Let's look at Michael. Michael's doctor has warned him, "Watch your fat and sodium intake!" He's also trying to lose a couple of pounds, so it won't hurt to compare nutrition facts. And that's exactly what Michael is doing. He loves soup, and he's trying to decide between two brands.

First, Michael needs to make sure he's comparing tomatoes with tomatoes, so he checks the serving sizes. One serving of Lovely Lentils is 1 cup, and one serving of Barley and Beef is 8 ounces. Are these equivalent? Yes. There are 8 ounces in 1 cup, so Michael (thankfully) doesn't need to do any extra calculations.

Now he needs to consider the fat. What is the percent of calories derived from fat for each soup? To find this, he needs to divide the number of calories from fat by the total number of calories.

Lovely Lentils: $90 \div 210 = 43\%$

Barley and Beef: $261 \div 320 = 82\%$

Whoa! Barley and Beef's percent of fat calories is twice as large as that for Lovely Lentils. There's a clear winner here.

Still, Michael wants to look at the amount of sodium in each soup. Again, Lovely Lentils is on top, with a much smaller amount of sodium; and that brand is higher in vitamins A and B, calcium, and iron.

But there's another reason why Michael should take a closer look at Lovely Lentils: the total calories. If he heats this soup up for lunch, he'll eat fewer calories. Just what the doctor ordered.

Constants

In many of these formulas, there is something called the *constant*. The constant never changes. In the BMI formula, 703 is the constant.

$$BMI = \frac{703w}{h^2}$$

The BMR formulas also have constants, which are shown here.

$$BMR_{women} = 655 + 4.3w + 4.7h - 4.7a$$

$$BMR_{men} = 66 + 6.3w + 12.9h - 6.8a$$

No matter who uses the formulas, the constants always stay the same. When *you* use these formulas, you introduce another constant. In the case of the BMI and BMR formulas, your constant is your height. That's a number that probably won't change, although (if you're being honest) your weight and age will.

The Heart of the Matter

Let's say you want to eat that entire bag of chips at lunch. To balance that out—and not gain weight—you'll have to move your body. There's math involved there, too.

When your heart is working hard, so is your body. So, when you're exercising, you should know how fast your heart is beating. Too close to normal, and you're not exerting enough energy. Too fast, and you could be pushing it.

Your heart rate is measured in beats per minute. Calculating your heart rate is easy: Just find your pulse and count. You could count for an entire minute, but there is an easier way—start at 0 and then count for only 10 seconds. Now you have to do some math.

Let's say that you counted 11 heartbeats in 10 seconds. How many heartbeats would you have in 1 minute? All you need to do is multiply by 6.

$$11 \cdot 6 = 66$$

Why 6? Because there are 60 seconds in a minute, there are six 10-second periods in a minute. (Here's another way to think of it: $60 \div 10 = 6$. And here's another: $10 + 10 + 10 + 10 + 10 + 10 = 60$.) So you need to multiply the number of times your heart beats during each 10-second period (which is 11) by 6.

This result is called your *resting heart rate*, or RHR. Another important number is your *maximum heart rate*, or MHR. This is the fastest your heart should beat, and it is not advisable for it to exceed that rate. You're not necessarily going to keel over if your heart rate reaches its maximum. But you shouldn't exercise for very long when your heart is beating that fast.

There are about half a dozen ways to calculate your maximum heart rate, and all of them come from highly respected experts. But for most folks, a simple formula works just fine:

MHR – 220 – *age*

Let's say that you're 40 years old. In that case, your MHR should be 220 – 40 = 180 beats per minute.

Your RHR and MHR are the bases of the other heart rate zones. And your heart rate zones will help you exercise most efficiently. The American Council on Exercise defines these zones this way:

Intensity of Exercise	Zone
Light to moderate exercise	55–65%
Moderate to vigorous exercise	65–85%
Very vigorous exercise	85%–MHR

Take a look at the table. Should you simply take the percent of your resting heart rate to find how fast your heart should be beating during this kind of exercise? No, because that would put your rate during exercise at *less than* your RHR. That doesn't make sense at all! In fact, there is a formula you can use to find your target heart rate in each of these zones.

$(M-R)p=z$

$z+R=Z$

M is maximum heart rate

R is resting heart rate

p is the percent from the Intensity of Exercise table

z is the zone

Z is the zoned heart rate

Let's take a look at an example. Jesse is a competitive sprinter. In order to stay on top of his game, he needs to monitor his heart rate during his daily workout. He's 25 years old, and his RHR (resting heart rate) is 72 beats per minute (bpm). He has to use a formula to find his MHR:

MHR = 220 – age

MHR = 220 – 25

MHR = 195 bpm

Jesse's trainer does the math and comes up with this:

- Jesse's heart rate should not exceed 195 bpm at any time during his workout.

- When he's warming up, his heart rate should be between 140 and 152 bpm.
- When he's in his regular workout period, his rate should be between 152 and 177 bpm.
- And when he does short, intense bursts—such as sprints—his heart rate should be between 177 and 195 bpm.

High-Intensity Interval Training

Mazzy is tired of spending her mornings at the gym. She'd rather be snoozing in bed. Her trainer suggests high-intensity interval training (HIIT).

The idea is to switch up the intensity of your exercise—dramatically—which Mazzy's trainer says will make her workout more efficient. The warm-up and cooldown are typical of most exercise routines. But in between, she'll cycle through intense and easy exercise, forcing her heart and muscles to keep up with the sudden changes. Here's a sample workout:

- Slow walk (warm-up) for 3 minutes
- Fast run for 30 seconds
- Slow walk for 1 minute
- Fast run for 30 seconds
- Repeat previous "walk 1 minute, then run 30 seconds" set for 8 more cycles (9 cycles in all)
- Slow walk (cooldown) for 3 minutes

Sounds great, but will Mazzy have time to sleep in and hit the gym before work? Or could she fit it into her lunch hour? Mazzy needs to know how long the workout takes.

Warm-up = 3 minutes

Fast run = 0.5 minute

Walk/run cycles = 1.5 minutes • 9 cycles = 13.5 minutes

Cooldown = 3 minutes

Total = 3 + 0.5 + 13.5 + 3 = 20 minutes

Mazzy can do it! She sets her alarm and hits the pillow happy.

Pumping Iron

Many studies have shown that muscle burns more calories than fat. That idea boils down to an amazing weight-loss secret: adding muscle mass boosts metabolism. Building muscle offers other health benefits, too. Creaky joints? Muscle strength can help. Worried about osteoporosis? Adding muscle can increase bone density and lower your risk of fractures. Want to reduce your chance of a heart attack? When the body is leaner, your ticker is healthier.

Whether you're using free weights or a machine at the gym, there are two parts of strength training: the amount of weight you're lifting, and the number of times you repeat each exercise.

A little bit of math can keep you from being a complete dumbbell at the gym. Take a look at this sample bench-press workout.

BENCH-PRESS WORKOUT

	Week 1	Week 2	Week 3	Week 4
SET 1	65% • 5	70% • 3	75% • 5	40% • 5
SET 2	75% • 5	80% • 3	85% • 3	50% • 5
SET 3	85% • 5+	90% • 3+	95% • 1+	60% • 5

A *set* is the number of times you lift a particular weight in a row. So each time you work out, you'll do three sets. And as the weeks progress, you'll change the number of times you lift that weight.

How do you know how much weight you should be lifting for each exercise? The chart tells you that, too. But you need some additional information. That's where something called the "1 rep max" comes in.

Also called 1RM, the 1 rep max is the most you can lift in 1 repetition. The percents in the chart refer to the percent of 1RM that you'll lift for each set. So if you can lift 10 pounds as your 1RM, then 85% of that would be 8½ pounds.

As the weeks progress, the sets get more and more intense. You start out lifting 65% of your 1RM, and by the third week, you're lifting 95% of your 1RM. (The last week is a resting week.)

So how do you find your 1RM? You choose a weight to lift for a particular exercise. Do the exercise with that weight. Count how many times you can lift the weight before your muscles are completely fatigued. Then use the Brzycki formula (a formula used to calculate your 1 rep max) to find your 1RM. Here it is:

$$1RM = w \cdot \frac{36}{37 - r}$$

w is weight

r is the number of repetitions

Let's try it out. Matt can bench-press 100 pounds five times before his muscles are fatigued. What is his 1RM?

$w = 100$

$r = 5$

$$1RM = 100 \cdot \frac{36}{37 - 5}$$

$$1RM = 100 \cdot \frac{36}{32}$$

$1RM = 100 \cdot 1.125$

$1RM = 112.5$

Matt's 1RM is 112.5 pounds. His personal trainer has suggested that he follow this workout in the first week:

Set 1	65% of 1RM, 5 reps
Set 2	75% of 1RM, 5 reps
Set 3	85% of 1RM, 5 reps

How much will he need to lift for each set?

Set 1	65% of 112.5
	0.65 · 112.5
	73.125
	Round to 73 pounds
Set 2	75% of 112.5
	0.75 · 112.5
	84.375
	Round to 84 pounds
Set 3	85% of 112.5
	0.85 · 112.5
	95.625
	Round to 96 pounds

Once Matt finishes the first month of his workout, he'll need to find his 1RM again. He has gotten stronger, after all! Once he has a new 1RM, he'll adjust his workout, this time adding more weight.

Weighing the Options

Matt has to add weights to the barbell to make it weigh the right amount: 73 pounds, 84 pounds, and so on. But the bar itself weighs something. If he slides 73 pounds of weights onto it, he'll have a barbell that weighs a lot more than 73 pounds. He has to consider how much the bar weighs. A standard Olympic bar weighs 45 pounds, so how much will Matt need to add to get to 73 pounds?

$73 - 45 = 28$ pounds

Here's another question to consider: Should Matt add 28 pounds to each side of the bar? Nope. He needs to put half of the 28 pounds on one side of the bar, and the other half on the other side.

$28 \div 2 = 14$ pounds on each side of the bar

The weights Matt is using are available in 50-, 25-, 10-, 5- and 1-pound sizes. What are his options for each side of the bar? He can use one 10-pound weight and four 1-pound weights; or two 5-pound weights and four 1-pound weights.

What if Matt doesn't have any 1-pound weights? In that case, he should probably choose one 10-pound weight and one 5-pound weight; or three 5-pound weights. How much more will Matt be lifting if he has no 1-pound weights? Just 2 pounds.

Glossary

401(k)

A program set up by a business to help its employees save for retirement. Contributions to a 401(k) are tax-deferred. Taxes on this money—and on the increases in the value of the account that occur through investments over the years—are paid only when the individual withdraws money from the account, which generally occurs after her or his retirement. Before that time, there are restrictions on when the individual can have access to the money in a 401(k) account.

Additive Identity

0; when you add zero to any number, the answer is identical to the number that you started with.

Adjustable-Rate Mortgage (or Variable-Rate Mortgage)

A mortgage agreement that sets an initial mortgage rate lower than that available on a fixed-rate mortgage. The rate can be adjusted periodically by the lender based on an index that is specified in the mortgage contract.

Algebra

The field of mathematics in which letters and other symbols are used to represent numbers.

Algorithm

A set of steps used to solve a problem.

Angle

A two-dimensional figure formed by two rays that share a common endpoint called the vertex.

Annuity

A savings program in which a fixed total amount of money is deposited into an account on a regular basis.

APR

Annual percentage rate, the stated interest on a loan for a 1-year period. When interest is compounded more than once a year, this number may be divided into a smaller amount for the interest rate used in an interest period. For example, if the APR is 6% but interest is charged monthly, the monthly interest rate is $6\% \div 12 = 0.5\%$.

Area

The amount of space occupied by a two-dimensional figure. Area is expressed in squared units.

Arithmetic

Basic operations (addition, subtraction, multiplication, and division) with whole numbers, fractions, decimals, and exponents.

Associative Property

A property of addition and multiplication. When you add or multiply more than two numbers, you can group the numbers together in any way you like without changing the answer. For example, $34 + (56 + 78)$ is the same as $(34 + 56) + 78$, and $2 \times (4 \times 6)$ is the same as $(2 \times 4) \times 6$.

Circumference

The distance around a circle or other closed curve.

Common Denominator

The same denominator in two or more fractions.

Commutative Property

A property of addition and multiplication. The order in which the numbers are added or multiplied does not change the answer. For example, $231 + 495$ is the same as $495 + 231$, and 23×41 is the same as 41×23.

Compound Discounts

More than one discount applied to a purchase. A typical example occurs when a customer uses a coupon for the purchase of something that the store is already offering at a sale price.

Compound Events

In a probability experiment, an outcome that is a result of a number of separate experiments. (Getting 3 Heads when four coins are tossed is one example; getting a sum of 7 when two dice are rolled is another.)

Compound Interest

Interest that is charged at regular intervals during the term of the loan.

Constant

A fixed value. Example: 10 is a constant in the expression $6x^2 + 10$.

Counting Numbers

1, 2, 3, 4, 5, 6, ...

Credit Score

A number ranging from 300 to 850 that is used as a gauge to determine whether a consumer has a history of repaying loans and the ability to do so at the present time. Lenders consider a person with a high credit score a good risk—that is, he or she is very likely to repay a loan.

Decimal Number

A number that includes a decimal point followed by digits. The digits to the right of the decimal point indicate values smaller than 1. Example: 5.7903

Denominator

The lower part of a fraction.

Difference
The answer in a subtraction problem. Example: The difference of 5 and 2 is 3.

Digit
A symbol used to make numerals. Example: 1, 4, and 7 are digits of 147.

Distributive Property
The complete name for this property is the Distributive Property of Multiplication over Addition. It states that if one number is to be multiplied by the sum of two or more numbers, then the same answer will be reached if the first number is multiplied by each of the numbers to be added and the sum of these products is found. For example, $34 \times (41 + 23 + 56) = 34 \times 41 + 34 \times 23 + 34 \times 56$.

Dividend
In a division problem, the amount being divided into parts. Example: In the problem, $20 \div 5 = 4$, 20 is the dividend.

Divisor
In a division problem, the number that is being divided into another number (called the dividend). Example: In the problem $20 \div 5 = 4$, 5 is the divisor.

Earned Income Credit
A refundable tax credit available to lower-income families and individuals.

Equation
A mathematical sentence that states equality. Examples: $4 + 2 = 6$ and $7x - 9 = 27$

Escrow
An account, separate from but associated with a mortgage account, into which a lending institution may require that a borrower make monthly deposits to ensure that money is available to pay property taxes and insurance premiums. An escrow account usually pays the borrower a nominal rate of interest.

Estimation
An educated guess, usually based on rounding the numbers before using any operations. An estimate is not an exact answer. Example: An estimate for $19 \bullet 4$ is 80.

Exponent
The exponent of a number tells how many times the number is to be multiplied by itself. Example: In $x5$, 5 is the exponent, and $x5 = x \bullet x \bullet x \bullet x \bullet x$.

Expression
Numbers, symbols, and operations grouped together. Example: $6x + 9$

Factors
Numbers that are multiplied together.

Fixed-Rate Mortgage
The interest rate on the mortgage loan agreed to at the time the papers for the mortgage are signed. It will be the rate of interest for the entire term of the loan.

Fraction

Part of a whole, written $\frac{a}{b}$, where a is any whole number, and b is any whole number. Examples: $\frac{2}{3}$ and $\frac{9}{1}$

Future Value of a Loan

The value of the loan at the end of the life of the loan. In a typical situation when the consumer borrows money, the future value will be 0 if the borrower takes the full time specified in the loan agreement to repay the loan. In a situation when the consumer is saving money, the future value is the amount of the money deposited by the consumer and all the interest earned on those deposits.

Greatest Common Factor (GCF)

The largest number that will divide evenly into two or more numbers. Example: 12 is the GCF of 60 and 24.

Improper Fraction

A fraction whose numerator is larger than its denominator. An improper fraction is always larger than 1. Improper fractions can be written as mixed numbers. Example: $\frac{10}{3}$

Integers

The set of numbers containing the whole numbers and their opposites: . . . , –5, –4, –3, –2, –1, 0, 1, 2, 3, 4, 5, ...

IRA

An individual retirement account (IRA) is an account in which money can accumulate on a tax-deferred basis until the individual is of retirement age. There are restrictions on the individual's access to this money before retirement. Taxes on it are paid as the individual eventually begins withdrawing money from the account.

Mean

The number found by adding all the data values in a set of numbers and dividing this sum by the number of data points. The mean is the measure of center that is most commonly associated with the word *average*.

Median

The number in the center of a set of data that have been organized in ascending or descending order. The median is occasionally—though less often than the mean—referred to as the average.

Mixed Number

A number that includes a whole number and a fraction. Mixed numbers can be written as improper fractions. Example: 4¾

Mortgage

A loan for the purpose of buying a home.

Multiple

The product of two numbers; also, a number that can be divided evenly by another number. Example: 25 is a multiple of 5.

Multiplicative Identity

1; when you multiply 1 by any number, the answer is identical to the number that you started with.

Number Sense

The ability to use and understand numbers. Number sense includes an understanding of number values, operations (including their properties), and estimation.

Numerator

The upper part of a fraction.

Operation

A mathematical calculation, most commonly addition, subtraction, multiplication, and division.

Order of Operations

The order in which calculations must be performed in an expression: parentheses, exponents, multiplication, division, addition, and subtraction.

Percent

One part of 100 represented with a percent sign (%). Example: More than 25% of the cars were red.

Percentage

A portion of the whole, usually stated in a general sense. Example: The percentage of red cars on the road has increased significantly over the last year.

Perimeter

The distance around a two-dimensional shape. The perimeter of a circle is called the circumference.

Periodic Rate of Interest

The rate of interest charged in each interest period.

Pi (π)

The ratio of the circumference of a circle to the length of the diameter of the circle. Pi is approximately equal to 3.14159.

PMI

Private mortgage insurance is a policy that a homeowner may be required to purchase by the company that lent the owner the mortgage money. This coverage ensures that the lending institution does not lose the mortgage money if the owner should default on the mortgage.

Points

A fee that a lending institution may charge the borrower at the time the papers are signed to complete a mortgage deal.

Present Value of a Loan

The value of the money in today's dollars. This number does not take into account the interest that may be charged for the use of these dollars during the life of the loan.

Probability

The study of determining the likelihood of outcomes. The term is also used to describe the ratio of favorable outcomes to all possible outcomes.

Product

The answer in a multiplication problem.

Quotient

The answer in a division problem. Example: In the problem $20 \div 5 = 4$, 4 is the quotient.

Range

In statistics, the spread of the values in a data set. The range is the difference between the largest and smallest numbers in the data set.

Ratio

A comparison of two numbers. Ratios can be written with a colon, a fraction line, or the word *to*. Examples: 2:1, $\frac{2}{1}$, and 2 to 1.

Remainder

The amount left over after division. Example: $15 \div 4 = 3$ with a remainder of 3. If a divisor is a factor of the dividend, the remainder is 0.

Rounding Off

Estimating a number to a specific place value.

Scientific Notation

A method for writing numbers in the form $x.xx \times 10^n$. This form is especially useful when writing numbers that are very large. For example, 1,234,567,891,234,567 is written as 1.23×10^{14} and 0.0000000000000123 is written as 1.23×10^{-14}).

Side

One of the line segments that defines a two-dimensional shape, or one of the surfaces that defines a three-dimensional shape.

Simple Event

In a probability experiment, an outcome that can occur in one way only. (Getting Heads when tossing a coin is one example; getting a 5 when rolling a die is another.)

Simple Interest

Interest that is charged once for the term of the loan.

Simplify

In mathematics, this means to put in the simplest form. Example: Simplify the fraction $\frac{5}{10}$. (Answer: ½) Example: Simplify the algebraic expression $2(x + 7)$. (Answer: $2x + 14$)

Square

Multiplying a number by itself. Example: $4^2 = 4 \cdot 4 = 16$

Standard Deviation

A number used to measure the average difference between the data points and the mean for a given set of data.

Statistics

The application of probability theory to analyze data. The term *descriptive statistics* is also used to describe a summary of data.

Sum

The answer in an addition problem. Example: The sum of 4 and 7 is 11.

Surface Area

The total area of a three-dimensional figure, such as a cube, cylinder, or prism.

Term

Part of an algebraic expression that includes a number, a variable, or the product of numbers and variables. Terms are separated by addition or subtraction symbols. Example: The terms of $6x2 + 4x -2$ are $6x2$, $4x$, and 2.

Unit Price

The cost for 1 unit. For example, $3.78 per gallon, $3.99 per pound, and $1.10 each (per unit) are unit costs. A means for comparing the costs per unit of a product that is offered for sale in packages that differ in size.

Variable

A symbol for an unknown number. A variable is usually a letter. Example: The variable in $2x + 9$ is x.

Vertex

A point where two or more line segments meet. The plural of vertex is vertices. Angles, squares, triangles, rectangles, cubes, prisms, and pyramids have vertices.

Volume

The amount of three-dimensional space an object occupies. Volume is expressed in cubic units.

W-2 Tax Form

A form used to provide a summary of the income earned and taxes paid by an employee during a year.

W-4 Tax Form

A form used to determine the amount of taxes that will be withheld from an individual employee's paycheck.

Whole Numbers

0, 1, 2, 3, 4, 5, 6, ...

Index